CHINESE TANKS & AFVs
1950–Present

CHINESE TANKS & AFVs 1950–Present

MARTIN J. DOUGHERTY

amber BOOKS

This Amber edition first published in 2019

Published by Amber Books Ltd
United House
London N7 9DP
United Kingdom
www.amberbooks.co.uk
Instagram: amberbooksltd
Facebook: www.facebook.com/amberbooks
Twitter: @amberbooks

ISBN: 978-1-78274-868-7

Editor: Michael Spilling
Designer: Andrew Easton
Picture research: Terry Forshaw

Printed in the UAE

Contents

Introduction

As China transformed itself into a global superpower, it needed to develop the capability to construct its own weaponry that could take on the best that any potential opponent might deploy.

Initial close relations with Soviet Russia meant that equipment could easily be obtained and locally produced copies made. However, differences of opinion over how best to practise Communism resulted in relations between Moscow and Beijing rapidly cooling from 1956 onwards.

Despite this, mid-twentieth-century Chinese military equipment closely resembled Soviet hardware – to the point where many Chinese-built weapons were direct copies of their Soviet equivalents. Although design philosophies gradually diverged, this was a slow process. Equipment bought overseas was suited to the doctrines of the provider, so to be effective the Chinese military had to embrace similar practices.

Newly designed weaponry had to fit with the adopted doctrine, so could not be radically different. Engineering habits, stockpiles of spares and supplies and the desire to avoid having to retool factories all guided Chinese weapons development in the years following the Sino–Soviet split.

Since the 1960s, the Chinese military has charted its own path, at times producing highly innovative equipment and at others finding similar solutions to the same problems as its rivals overseas. Today's equipment is the result of that process, and of China's status as a superpower. As such, its forces might be expected to take on a range of challenges

A Chinese ZBD-03 IFV participates in the 2016 International Army Games organised at the Rayevsky training ground, Novorossiysk, Russia.

A Chinese Type 96 second-generation MBT undergoes training in 2017. The Type 96 has remained a frontline tank since its introduction in 1997.

anywhere in the world, ranging from a low-level counter-insurgency campaign or providing protection to a humanitarian assistance operation right up to fighting a major war against a technologically advanced opponent.

Appraisals of Chinese tank strengths vary considerably, with estimates ranging from 5–7,000. Of these, roughly half appear to be modern third-generation designs, with the remainder made up of modernised second- and first-generation tanks.

Growing capability

Armoured vehicles are the spearhead of that capability, whether their role is direct combat, to transport, protect and support infantry, or to carry supporting systems such as artillery weapons and air defences. They operate in an environment where hazards come in many shapes and forms, with new threats and innovations on old ones constantly appearing. This requires ongoing development, seeking to stay on the right side of the threat/defence equation.

Whether it is digital camouflage, improved automation or urban survival equipment, armoured vehicles continue to receive modifications that improve survivability, efficiency and fighting power. An obsolete tank is still a potent force in some environments; in others it is simply a mobile deathtrap. In a world of tight military budgets, the decision to upgrade a design or replace it with something a generation more advanced is not a simple one. It might be cost-effective to let an old design carry on, but eventually any military must face the hard truth that you fight with what you have. Chinese planners thus have to deal with the twin problems of needing enough armoured vehicles of adequate capability to equip their large forces, while also requiring vehicles with sufficient capabilities to deal with a high-end threat if one emerges.

TJ107

MAIN BATTLE TANKS

At the end of World War II, the People's Liberation Army (PLA) of China found itself in possession of a great deal of assorted military hardware. This hodgepodge of equipment had to serve as the nation reestablished its identity in the late 1940s, but the logistical burden alone meant that rationalization was desirable even if outright military efficiency had not been a consideration.

This chapter includes the following tanks:

- Type 58
- Type 59
- Type 62
- Type 63 amphibious tank
- Type 69
- Type 79
- Type 80
- Type 88
- Type 85
- Type 90
- Type 96
- Type 99

The Type 99 main battle tank, which equips the most prestigious armoured units of the People's Liberation Army (PLA), is the result of a long development process that began with Soviet-supplied T-34s in the 1950s.

Type 58

Type 58 was the (probably unofficial) designation given to ex-Soviet T-34 tanks in Chinese service.

Although the T-34 is widely considered to have been the best tank of World War II, it had many faults. These included a tendency for the transmission to break down, gearbox and visibility issues and a very cramped internal layout. These problems made T-34s individually less effective than they should have been. However, the T-34 did put an effective gun on the battlefield, gave it adequate mobility and protection and, perhaps most importantly, was available in the huge numbers required.

Some of the T-34's defects had been ironed out by the end of the war, and on balance it was a cost-effective solution to the problem of fielding large numbers of effective tanks. Better vehicles were available to the Soviet Union by then, but huge numbers of T-34s had been built during the war. The ideal solution was to find an overseas buyer and replace the T-34 with vehicles more suited to a potential conflict with the Western nations.

At the same time, China had little or no capability to produce armoured vehicles, and needed a great many. During 1950–55, hundreds of T-34s and smaller numbers of other tanks were transferred from the Soviet Union to the People's Republic of China. The Type 58 designation may have arisen from plans to begin indigenous production of a licence-built T-34 in 1958. However, it is uncertain whether any such plan ever existed. At this point, the basic design was almost 20 years old, and while the original 76mm (3in) gun had long since been replaced with a more effective 85mm (3.34in) weapon, some defects of the original remained.

The T-34's Christie suspension, while theoretically giving a good road

Type 58

Date 1950–55

Number Produced Probably around 2000 delivered from the Soviet Union

Weight 30.2 tonnes (29.7 tons)

Length (with gun) 8.15m (26ft 9in)

Width 3m (9ft 10in)

Height 2.6m (8ft 6in)

Engine V12 diesel, 372kW (493hp)

Maximum Road Speed 38km/h (26mph)

Crew 4

Armour 30–80mm (1.18–3.15in)

Main Armament 85mm (3.34in) ZIS-S-53 or D5-T

Main Gun Ammunition Stowed 55

Type 58

T-34s served in the Korean conflict, crewed by Chinese and Korean personnel. Claims of their exploits have been greatly inflated, though by all reports the T-34 gave good service.

Type 58-I

Most of the evidence for the Type 58-I upgrade package comes from photographs rather than official specifications, making it difficult to ascertain exactly what modifications were made.

speed, was outdated and considered by many experts to be a poor solution to the problem of tank mobility. It also used up a lot of internal space in an already cramped design. The gearbox and transmission, even when not suffering from reliability issues, made the tank tiring to drive. Performance was often limited by the driver's inability to force the vehicle into a higher gear.

Outdated model

The T-34/Type 58 could be a deathtrap for its crews. Its armour was prone to spalling when a round failed to penetrate, and if a hit did penetrate, there were fuel tanks in the fighting compartment, which posed a severe hazard to the crew. Escape was also difficult if the tank was disabled, all of which contributed to poor crew survivability rates.

Despite these problems, the Type 58 provided China with an armoured warfare capability that had not existed before. By the time China was in a position to produce its own tanks, the T-34/Type 58 design was outdated. Many components of the Type 58 were put into production, but this may have been intended to support the purchased tanks rather than to create new ones. It is possible that there was a serious intent to produce new-build Type 58s, but in any case the much more modern Type 59 was chosen as China's first indigenous armoured fighting vehicle (AFV). Sources conflict, but it appears that the 'Type 58' designation was unofficial and T-34s in Chinese service were simply referred to by their Soviet designation.

Type 58s in Chinese service suffered from many of the same issues as Soviet equivalents, though some defects such as a lack of all-round vision were addressed. The solution to this problem was rather basic, consisting of a cupola that on some examples might lack any sort of glazing – armoured or otherwise – in the vision slits. Likewise, attempts were made to strengthen the transmission and thereby improve reliability. Had a production version ever materialized, it seems likely that these fixes would have been better integrated, but by the time the Type 58 entered Chinese service better options were available.

All of the T-34s received by China were the later T-34/85 model, with an 85mm (3.34in) gun. North Korea, among other clients, received earlier versions equipped with the original 76mm (3in) gun. These proved devastatingly effective against the

A Chinese T-34/85 '406' in the Beijing Tank Museum. To the left are two IS-2 heavy tanks in Chinese markings.

South Korean military, which had few anti-tank weapons, but did not fare as well once post-war Western tanks began to arrive in the conflict.

Nevertheless, the obsolescent T-34 proved that it was still a viable weapon on the battlefield of the 1950s. On balance, the T-34/85 was a good buy for the fledgling Chinese military. Slightly modified to create the version generally known as the Type 58, it created a ready-made capability that was not too far behind the times to be effective, and allowed experience to be gained of maintaining and operating an armoured force.

Type 58-II

As with many Chinese designs, the Type 58-II designation was arbitrarily imposed by Western observers who had only partial information and may have imagined distinctions that did not matter within Chinese service. This model appears to have a Soviet DShK 1938 12.7mm (0.5in) mounted on the turret.

Type 59

Like the Type 58, the Type 59 was based on a Soviet design. It was derived from the T-54, which was developed as a replacement for the T-34.

The Type 59 was the first 'modern' Chinese tank, and also the last to be produced with direct Soviet assistance.

Whereas the T-34 was crashed into production at a time when the very survival of the Soviet Union was in doubt, the T-54 began its design process in 1944. By this time, the tide of World War II had turned, and Soviet designers were able to concentrate on creating a replacement for the T-34 that offered significant improvements.

Perhaps most importantly, the T-54 did away with the T-34's notoriously unreliable transmission. It also had a new engine and a 100mm (3.9in) gun. This was an extremely potent weapon for the time, capable of penetrating any known tank. Armour protection was also improved, while the general design was optimized for the Soviet style of armoured warfare.

Whereas Western tanks tended to emphasize quality over quantity and were designed to enhance survivability

Type 59
Date 1958
Number Produced 10,000+
Weight 36 tonnes (35.4 tons)
Length (with gun) 9m (29ft 5in)
Width 3.27m (10ft 7in)
Height 2.59m (8ft 5in)
Engine 12150L V-12 diesel, 388kW (520hp)
Maximum Road Speed 50km/h (31mph)
Crew 4
Armour Min–max RHA 20–203mm (0.8–8in)
Main Armament 100mm (3.9in) Type-59 L/53.5 rifled cannon
Main Gun Ammunition Stowed 34

Type 59 (early production)
Early Type 59s were more or less direct copies of the Soviet T-54. Among their drawbacks was a non-rotating turret floor, requiring an alert crew to avoid injury when traversing the gun.

Type 59-I

The addition of side skirts changed the appearance of the Type 58-I considerably, but its most important improvement was a new gun supported by a laser rangefinder.

Type 59-II

The Type 58-II also received a new gun, in this case a Western 105mm (4.1in) rifled weapon. This was the first use of NATO ammunition by the Chinese armed forces.

Type 59-II
Date 1982
Number Produced 448
Weight 36.5 tonnes (35.9 tons)
Length (with gun) 9.4m (30ft 9in)
Width 3.27m (10ft 7in)
Height 2.59m (8ft 5in)
Engine Model 12150L V-12 diesel, 388kW (520hp)
Maximum Road Speed 50km/h (31mph)
Crew 4
Armour Min–max RHA 20–203mm (0.8–8in) + composite + ERA
Main Armament 105mm (4.1in) Type 81 rifled cannon
Main Gun Ammunition Stowed 34

by sniping from hull-down defensive positions, Soviet tanks were built with a different doctrine in mind. With a low silhouette and a rounded turret to deflect marginal hits, Soviet tanks were optimized for a fast-moving advance by large numbers of armoured vehicles.

Simplified copy

Developed versions of the T-54 appeared throughout the late 1940s and into the 1950s. These were observed with interest by the Chinese military, which selected the tank for indigenous production. Relations between Moscow and Beijing were at this time still cordial, permitting an agreement to allow licence-building of a T-54 variant and Soviet assistance in setting up the production programme. Designated the Type 59, the new tank was almost a direct copy of the Soviet original, but was simplified in order to streamline production. The T-54 had received a gun stabilization system in 1957, but this was not included in the Chinese version. Nor was the T-54's infrared searchlight or nuclear, biological, chemical (NBC) protection system, which at the time were not deemed necessary or cost-effective.

This simplification meant that early Type 59s could not fire accurately on the move or operate in darkness, but these problems were rectified on the later Type 59-I model. Further upgrades continued throughout the Type 59's long career.

Although not as technically sophisticated as contemporary Western tanks, the Type 59 was a potent vehicle offering excellent off-road mobility and vastly improved reliability over

Type 59-IIA

The Type 59-IIA gained composite armour and an array of equipment upgrades. Experience gained with this model was useful in planning future composite-armoured AFVs.

Type 59-IIA

The Type 59-IIA received further upgrades in the 1990s (such as this model here), including improved fire control and electronics, to ensure it remained effective in the modern battlefield environment.

Type 58 IIA

The thermal sleeve on this mid-2000s Type 59 would have been irrelevant with the original design's basic sights, but with modern fire control the small variances due to temperature make a significant difference.

Type 59 tanks move through a snowy landscape, December 2017. It is estimated that as many as 2300 upgraded Type 59s are still in Chinese service. There are many situations where an old tank with good fire control is entirely adequate.

its predecessor. At the time of its introduction, the Type 59's gun could penetrate anything it could hit, although the design had some tactical limitations.

In keeping with Soviet practice, the Type 59 had a low silhouette at the price of limited gun depression. This was no real drawback during a fluid battle, but on the defensive the Type 59 would be forced to expose more of its hull than a Western tank of the same era.

A 'hull-down' firing position could be established by finding an obstruction at just the right height and hiding behind it, but in most cases the tank would have to drive onto a slope that would raise the bow. A high turret and large angle of gun depression allowed a tank to position itself with nothing but the turret and gun showing and the hull protected by the obstruction in front. The Type 59 could take advantage of this situation to only a limited degree.

Rapidly cooling relations between China and the Soviet Union eventually led to war in 1969. Soviet assistance with tank production had long since been curtailed by that point, and the Type 59 was further developed by Chinese designers without outside help. However, Soviet armoured fighting vehicle design did continue to influence the Type 59. Studies of a T-62 captured from the Soviet Union in 1969 proved useful during the later development of the Type 59 and its derivatives.

Much like its Soviet counterpart, the Type 59 achieved good success on the international market. Its balance of capabilities versus simple design and relatively low cost made it attractive

Type 59G

The Type 59G incorporates many features found on third-generation MBTs – such as angular modular armoured tiles on the turret and ERA blocks – in a package cheap enough to be attractive on the international export market.

Type 59G
Date 1999
Number Produced 2312
Weight 40 tonnes (39.4 tons)
Length (with gun) 9.1m (29ft 9in)
Width 3.27m (10ft 7in)
Height 2.59m (8ft 5in)
Engine 12150L7 V-12 diesel, 544kW (730hp) or 566kW (800hp)
Maximum Road Speed 50km/h (31mph)
Crew 4
Armour RHA + composite + ERA + side skirts
Main Armament 125mm (4.9in) L/55 smoothbore cannon
Main Gun Ammunition Stowed 34

to many buyers, and examples are still in service today. More importantly, the Type 59 launched China's first generation of main battle tanks (MBTs) and was the basis for a number of other designs. Some served in the MBT role while others fulfilled specialist tasks. The Type 59 and its immediate derivatives formed the backbone of China's armoured forces for many years, with significant numbers remaining in service or held in reserve to the present day.

In combat

During the Cold War, any clash between armies supplied by Eastern and Western powers was observed with great interest. Wars in Asia and the Middle East were a proving ground for the designs and the doctrines that might one day meet in a Third World War. Thus, the deployment of Chinese Type 59s in Vietnam in the late 1960s permitted strategists and tank designers the world over to see how the Type 59 fared against a similar generation of Western armoured fighting vehicles.

China sent Type 59s to assist North Vietnam, which also fielded the almost identical T-54 as supplied by the Soviet Union. Their chief opponents were American M48 MBTs and the light M41 Walker Bulldog, but the conflict also permitted experience to be gained of armoured operations in general. The performance of the Type 59 was considered acceptable, though tank-versus-tank actions were very rare due to the Vietnamese terrain and the nature of the war in general.

The Type 59 saw action in Chinese hands during the Sino–Vietnamese war of 1979, although Chinese tanks in general did not perform well in the difficult terrain. Examples exported overseas were later involved in numerous conflicts. In many cases, these were local variants or upgraded models, which remained at least borderline effective despite the age of the basic design.

Modern reserve

Today, the Type 59 serves in several African militaries and remains an important part of the Chinese armoured forces. With such a large military to equip, China chose to upgrade and keep the ageing Type 59 – officially designated WZ-120 – in service rather than replacing it wholesale. This represents a 'high-end/low-end' attitude to equipment, whereby budget funds freed up by not having to replace thousands of tanks can be funnelled into developing very high-capability vehicles to equip spearhead formations.

There are many situations in the modern world where 'any tank is enough tank', and the Type 59 can serve almost as well as a far more expensive top-end design. Even where capability is more important, the Type 59 was designed for massed tank warfare and can still fulfil this role well enough. Cynical observers might also wonder if vast numbers of modestly capable Type 59s might act as a shield and concealment for the PLA's top-end vehicles, but it seems likely that the decision to keep them in service was more pragmatic. The Type 59 can still get the job done, and replacing it with something better would not provide the returns its enormous cost would entail. Thus it is likely that this venerable design will continue to receive minor upgrades and remain in service for some years to come.

Although heavily modified, the Type 59 has a heritage stretching back to the final days of World War II. It was designed in an era where massed tank warfare was the expected future of conflict.

Type 62

The terrain in much of China is unsuited to tank operations. As a result of operating there during the 1930s, the Japanese army decided to continue building very light tanks even though more survivable options were now available.

This resulted in Japanese armoured forces being greatly overmatched by their Allied equivalents, and vulnerable to even obsolete anti-tank weapons. It seems that Japan learned the wrong lessons from operating in China. Japanese experiences in World War II may have influenced Chinese thinking when the time came to produce their own light tank.

As China developed its military capabilities, it needed a tank that could operate effectively in this environment without being useless elsewhere. It was thus necessary to ensure the design did not sacrifice too much fighting power in return for good rugged-terrain performance. The result was the Type 62 light tank, officially designated WZ131, which was derived from the Type 59 by way of some significant modifications. The general lines were similar and obviously 'Soviet' in origin, with a

Type 62
Date 1963
Number Produced 1500+
Weight 21 tonnes (20.6 tons)
Length (with gun) 7.9m (25ft 9in)
Width 2.86m (9ft 4in)
Height 2.25m (7ft 4in)
Engine 12150L3 V12 diesel 321kW (430hp)
Maximum Road Speed 60km/h (40mph)
Crew 4
Armour (min–max) 15–50mm
Main Armament 85mm (3.34in) Type 62-85TC rifled gun
Main Gun Ammunition Stowed 47 rounds

Type 62
The Type 62 was visually similar to the Type 59, and most easily recognized by its 85mm (3.34in) gun, which is shorter than the weapons used by the Type 59.

Type 62
A PLA Type 62 in North Vietnam, Sino–Vietnamese War, 1979. The Type 62 performed well enough in the terrain of Vietnam, but losses to RPGs prompt an upgrade package.

Type 62-I (Early Production)

The Type 62-I gained additional protection at the expense of increased weight, along with other improvements, including a laser rangefinder.

rounded turret and low silhouette, but the smaller hull permitted a much lower overall weight.

Size reduction was achieved by moving the first and second of the five road wheels closer together and using lighter construction. This might have weakened the suspension and tracks, but the load on them was also greatly reduced. The three-man turret, positioned further forward than on the Type 59, carried an 85mm (3.34in) gun – shorter and much lighter than the Type 59's 100mm (3.9in) weapon – and the armour was much thinner as well.

Light tank

Weighing in at 21.3 tonnes (21 tons), less than two-thirds the weight of a Type 59, the Type 62 had lower ground pressure despite narrower tracks and could maintain higher speeds. This provided some protection, as did the ability to create a smokescreen by injecting fuel into the exhaust, but during the 1979 Sino–Vietnamese conflict the Type 62 proved vulnerable to infantry anti-tank weapons such as the ubiquitous RPG-7.

The initial electronics fit on the Type 62 was basic in order to keep costs down, but upgrades began to appear along with survivability modifications. It is not clear how uniform or official some of these upgrades were, but some Type 62s appear to have gained additional armour along with formally mandated enhancements such as a laser rangefinder for the main gun.

Secondary armament

Like the Type 59, the Type 62 boasted a significant secondary armament. In addition to the usual co-axial 7.62mm (0.3in) machine gun and a 12.7mm (0.5in) heavy machine gun mounted on the turret top, the Type 62 also mounted a 7.62mm (0.3in) bow machine gun. Common during World War II, this practice has rarely been seen since. Another machine gun could be fitted on the turret top for use by the loader against aircraft or infantry threats.

Many examples were seen sporting side skirts. First developed during World War II, skirts offered some protection against shaped-charge high explosive anti-tank (HEAT) weapons such as RPG-7 projectiles. Too thin to be considered armour as such, skirts are sufficient to cause a HEAT round to detonate, ensuring that the plasma jet formed by the warhead focuses on a point

Type 62-I (Late Production)

The vulnerability of the laser rangefinder in its position above the gun barrel is obvious on this 1990s-era Type 62-I.

short of the tank's armour. Although a specialized defence, skirts helped to counter the primary threat to a light tank: relatively light weapons carried by infantry.

Type 62 tanks were sent to assist North Vietnamese forces in the 1960s and early 1970s, and later fought against them in the Sino–Vietnamese war of 1979. Although better suited to the terrain of Vietnam than the heavier Type 59, numerous Type 62s were disabled by infantry-launched RPGs, leading to an enhanced-survivability programme.

The Type 62-I, officially designated WZ131A, appeared in 1979, with further modifications continuing over the next decade. Another upgrade programme created the Type 62G from 2000 onwards. This version featured a new welded turret to replace the original cast one and contained a 105mm (4.1in) gun. Upgraded electronics were also received. However, by this time the Type 62 had been phased out of frontline service with the Chinese military.

Export success

The Type 62 found favour on the international market as a simple and cheap armoured vehicle well suited to operations in rugged terrain. Some examples still serve in Africa

Type 62G

Date 2000
Number Produced Not Known
Weight 21 tonnes (20.6 tons)
Length (not including gun) 7.9m (25ft 9in)
Width 2.86m (9ft 4in)
Height 2.25m (7ft 4in)
Engine 12150L3 V12 diesel 321kW (430hp)
Maximum Road Speed 60km/h (40mph)
Crew 4
Armour (min–max) 15–50mm
Main Armament 105mm (4.1in) rifled gun
Main Gun Ammunition Stowed 47 rounds

Type 62G

By the mid-2000s, the small numbers of Type 62s remaining in Chinese service were assigned to reconnaissance units supporting artillery formations. The longer 105mm (4.1in) gun is obvious in this illustration.

and Asia, including original-design Type 62s as well as upgraded models. A variant, designated Type 70, was created by the addition of gun stabilization and improved fire control. The Type 70 may or may not have received additional or modified armour protection as well. If it did, there are few visual indications.

Variants

The chassis was used as the basis for variants, including a self-propelled gun as well as armoured recovery and engineering vehicles. With many components in common with the rest of the Type 59 family, such vehicles were cost-effective to produce and maintain. Some overseas customers created their own modified versions, notably self-propelled guns and armoured personnel carriers.

The Type 62 provided a useful armoured capability for army and marine units operating in areas where heavier tanks were ineffective. This was not restricted to rough terrain; the Type 62 could cross weak bridges in underdeveloped areas, maintaining mobility without having to ford a watercourse or bring up bridging equipment.

Although the Type 62 is no longer considered viable by Chinese forces, the light tank concept remains valid and new variations on the theme have continued to appear.

The Type 62 was not, broadly, a success story but the Chinese armed forces still consider light tanks to be a useful part of their arsenal.

Type 63 Amphibious Tank

Although China probably did not specifically create the Type 63 for the purpose, an amphibious assault platform would be necessary for an invasion of Taiwan.

Although unable to take on battle tanks, an amphibious light tank offers many advantages beyond the capability to support a beach landing. The ability to quickly cross water obstacles or move through wetlands enables such vehicles to maintain an advance where a heavier force would need to stop and prepare a crossing. Amphibious vehicles can also be pushed forward across the obstacle to protect such an endeavour. A tank light enough to be made amphibious is also able to operate in rugged terrain or where a low ground pressure is useful, such as snow or marshy conditions.

Both the People's Republic of China and Taiwan – last refuge of the Nationalist faction defeated in the Chinese Civil War and officially known as the Republic of China – claim to be the legitimate government of all China. The level of tension varies over time, but the possibility of war will always exist. If that happened, an invasion would be vigorously defended by Taiwan and probably its allies, and armoured support would be essential in getting troops ashore.

The problems inherent in creating a suitable vehicle are not minor. Buoyancy can only be achieved by means of enclosed volume, but increasing size drives up weight and reduces the maximum possible armour thickness that can be applied. Weaponry is also a consideration; to be effective as a tank, or at least a direct-fire support platform, a vehicle needs not only a reasonably sized gun but also a turret to traverse and protect it, along with sufficient ammunition for an engagement. Even a modest increase in gun calibre drives up the weight of all these items, in turn requiring an increase in hull size and more overall weight.

The Chinese solution to these problems was visually very similar to the Soviet PT-76 amphibious light

Type 63-II Amphibious Tank

The Type 63 has obvious similarities with the Russian PT-76, but mounts a more powerful gun on a larger hull.

Type 63-II (WZ211-2)
Date 1963
Number Produced 1550
Weight 19.83 tonnes (19.5 tons)
Length (with gun) 8.44m (27ft 8in)
Width 3.2m (10ft 6in)
Height 2.52m (8ft 3in)
Engine 12150L-2 diesel 300kW (402hp)
Maximum Road Speed 64km/h (40mph)
Crew 4
Armour (min–max) 10–14mm (0.39–0.55in)
Main Armament 85mm (3.34in) Type 62-85TC rifled gun
Main Gun Ammunition Stowed 47 rounds

Type 63-II Amphibious Tank

The Type 63 and vehicles like it are highly useful to the Chinese marine forces as mobile fire support platforms.

Type 63-II Amphibious Tank

The Chinese armed forces have a long tradition of experimenting with camouflage systems to suit the many different conditions they must operate in.

Type 63-II Amphibious Tank

Like the two Type 63-IIs above, this example belongs to the marines. It is wearing a three-tone camouflage pattern from the 1990s.

Similarities to the Russian PT-76 are obvious, but the Type 63 is very much a Chinese design. Its turret was originally developed for the Type 62 light tank.

tank. It had the same large, angular hull and rounded turret. This was not coincidental; the Soviets had already fielded a vehicle that fulfilled the amphibious tank role, so starting completely from scratch would be a waste of effort. A number of PT-76s were available for study during the design process. However, the Type 60, as the original design was called, was not a direct copy. It was a little larger than the Soviet version with better buoyancy, and mounted a gun of greater calibre.

Prototype

The prototype, delivered in 1959, was rejected. Among its faults was a lack of power from the 179kW (240hp) engine. This was comparable with the powerplant of the lighter Soviet PT-76, which was also upgraded later in the vehicle's career. A 299kW (400hp) engine provided much better land mobility and swimming performance.

After more work, the design was adopted in 1963 and thus became generally known as Type 63. It used the lightweight turret from the Type 62 light tank, complete with 85mm (3.34in) gun and coaxial 7.62mm (0.3in) machine gun. The small size of this turret limited gun depression to four degrees, but this was less of a problem in a vehicle that would, by virtue of its role, spend much of its time lower than its targets.

Preparation for swimming involved starting internal pumps and raising the trim vane at the front of the hull, which prevented the front deck and the driver's periscopes from being flooded. Propulsion in water used twin waterjets – the same method as the Soviet PT-76 – but the Type 63 was a little faster on water and significantly more so on land.

Mounting an 85mm (3.34in) gun, the Type 63 proved to be an adequate fire support platform, but the compromises necessary in its design meant it was protected only against small-arms fire. There are many situations in which this is enough, but the combination of light protection, large size and its position at the head of an amphibious operation have ensured that the Type 63 has always suffered heavy casualties in action.

Type 63A

The original version had minimal electronics, making the gun inaccurate even if it had been stabilized. There was no NBC protection system, either. Upgraded versions appeared in the 1970s, gaining infrared sighting and a laser rangefinder. The Type 63A, appearing in the 1990s, was more than an upgrade package but incorporated significant redesign work. Although more seaworthy than the Soviet PT-76, the original Type 63 was designed with river crossings and short-range amphibious operations in mind. The new version was intended to make landings from several kilometres out to sea. An improved hull form and a new 432kW (580hp) engine, along with additional flotation tanks to improve stability, created a version that was little short of an entirely new vehicle.

In addition to enhanced swimming performance and mine protection, the Type 63A gained a new turret mounting a 105mm (4.1in) rifled gun. This weapon was similar to that used on other Chinese armoured vehicles, but with reduced recoil to avoid destabilizing the tank in water. The new gun also had the capability to launch guided missiles. Gun-launched missiles have a chequered history, but in theory offer accurate firepower even when the tank is afloat and can engage helicopters as well as ground targets.

There are few amphibious armoured vehicles on the international market, making the Type 63 one of the only options available. Overseas users included Vietnam, Pakistan, Cambodia and Venezuela, though not in any great numbers. Albania took delivery of 200, but these are no longer in service.

Despite its vulnerability to anything heavier than small arms, the Type 63 has served in a number of conflicts including the Vietnam War, the Sino–Vietnamese conflict, and the civil war in Sri Lanka. Its replacement, designated ZBD-2000, has a generally similar appearance and is little better protected, suggesting that the Type 63's designers did all they could to create an effective and survivable amphibious tank.

Type 63A Amphibious Tank

In addition to a redesigned hull, the Type 63A was given a new turret. Like the previous version this came from a non-amphibious light tank – the Type 62G.

Type 63A

Date 1997
Number Produced Not Known
Weight 22 tonnes (21.5 tons)
Length (with gun) 9.6m (31ft 6in)
Width 3.2m (10ft 6in)
Height 2.52m (8ft 3in)
Engine 12150L-2 diesel 432kW (580hp)
Maximum Road Speed 64km/h (40mph)
Crew 4
Armour (min–max) 10–14mm (0.39–0.55in), plus side skirts
Main Armament 105mm (4.1in) rifled gun; laser-guided ATGM
Main Gun Ammunition Stowed 47 rounds

Type 69

The Type 59 main battle tank was developed and put into production with Soviet assistance, but this was curtailed after 1960 as relations soured between Beijing and Moscow.

Development of the tank that would come to be known as the Type 69 went on in the shadow of a possible conflict with the Soviet Union and its Warsaw Pact allies, creating new requirements while at the same time permitting a move away from conformity with Soviet specifications and components.

Despite this, the Type 69 retained a distinctly 'Soviet derived' appearance, with five large roadwheels and a low hull topped by a rounded turret. The 100mm (3.9in) smoothbore gun was fitted with a basic stabilization system; along with an infrared searchlight and laser rangefinder, this improved targeting capabilities although gun depression was limited to five degrees. The provisional Type 69 was still less sophisticated than many of its contemporaries and could not shoot on the move with any real degree of accuracy.

The Type 69 had much the same dimensions and retained the same crew layout as its predecessor, with the driver in the hull and the commander, loader and gunner in the turret. Interior space remained limited, especially once NBC protection and additional electronics were fitted. Overall, it was very much a member of the first generation of Chinese main battle tanks and of the T54 family.

At the same time as the Type 69 was in development, the Soviet Union began producing the T-62 in an attempt to keep pace with Western tank design. An example was captured during the Sino–Soviet clash of 1969, permitting some of its advanced features to be reverse-engineered and incorporated into a Type 69-I prototype. This was developed to create the production model, which gained a stabilized 100mm (3.9in) rifled gun supported by much improved fire control.

Type 69-II

The Type 69-II entered service in 1982 and, like its predecessors, found favour on the international market. Although Chinese design philosophy had begun to diverge from its Soviet roots, Chinese tanks remained generally compatible with the T-54 family used

Type 69
Date 1982
Number Produced Up to 3200
Weight 35.6–37 tonnes (35–36.4 tons)
Length (with gun) 8.65m (28ft 4in)
Width 3.29m (10ft 8in)
Height 2.8m (9ft 2in)
Engine 12150L-7BW diesel 432kW (580hp)
Maximum Road Speed 50km/h (28mph)
Crew 4
Armour Up to 180mm (7in)
Main Armament Type 58 100mm (4.1in) rifled gun
Main Gun Ammunition Stowed Not Known

Type 69
The prototype Type 69 clearly showed its Soviet influences. These were only increased when a captured T-62 yielded ideas for improvements.

Type 69-I

The Type 69-I was produced in limited numbers to evaluate its new features. It was followed into service by the further upgraded Type 69-II.

by many nations. Similar component specifications simplified maintenance and reduced logistics overhead, while tanks of Chinese and Soviet origins were compatible in terms of training, doctrine and performance.

Both Iran and Iraq, opponents in the early 1980s, fielded Type 69 tanks, as did Pakistan, Bangladesh, Sri Lanka, Sudan and Zimbabwe. Iraq was the largest user outside China, though none of its Type 69s are still in service. Likewise, the Type 69 is no longer in Chinese service as a combat vehicle, though modified examples may still be encountered in specialist roles. Some non-Chinese users made local modifications to suit their own needs, though the extent of any alterations was limited by the available hull and turret space. Iraqi Type 69s gained a rifled 105mm (4.1in) gun of Western origin or a 125mm (4.9in) smoothbore weapon of Soviet design.

Armour protection was also enhanced in developed and modified versions. Notably, explosive reactive

Type 69 tanks take part in a military parade in Beijing. Cold War era Chinese tanks retained the features of their Soviet ancestors, including a low silhouette and rounded turret optimised for fast-moving massed tank assaults.

Type 69-II Prototype

The Type 69-II featured a new, more accurate rifled gun assisted by a dual-axis stabilisation system and improved fire control.

Type 69-IIIA

The Type 69-III or Type 79 marked a move away from Russian design features and increased use of Western technologies, such as its 105mm (4.1in) rifled gun.

armour (ERA) designed to disrupt a shaped-charge weapon was added to many Type 69s as part of both official and unofficial upgrade programmes. ERA could be applied in blocks over the top of existing armour, increasing weight but not requiring internal modifications.

Type 69-III

Now that China was no longer reliant on the Soviet Union for assistance, the option was available to use Western technology and equipment. The Type 69-III, also known as the Type 79, incorporated a 105mm (4.1in) gun derived from the British L7. This weapon was developed at a time when Soviet tanks were the most likely opponent for Western forces; its adoption suggests that Chinese military planners considered another clash with the Soviet Union likely. Even if that did not occur, Soviet armoured fighting vehicles were arming many nations around the world, so a gun developed to counter them might be a wise investment.

The Type 69 met frontline Western tanks during the 1991 Gulf War, and was greatly overmatched. It was not obsolete as such, having entered service around the same time as its opponents, but it was not in the same class. Defensively, the Type 69 was inadequate to cope with 120mm (4.7in) rounds from Western tanks, which were also far more likely to achieve a first-shot hit. This situation was even worse if the Type 69 was on the move. However, its design reflected a different philosophy. The Type 69 was designed to be cheap and to be fielded in large numbers. It was no great surprise that it was outmatched by vastly more expensive and sophisticated tanks designed for the primary purpose of killing cheaper ones.

Type 79

The Type 79 represented the final development of the Type 69, and thus the culmination of the first generation of Chinese MBTs.

It is sometimes designated Type 69-III, and may or may not be considered a separate model. It was, essentially, a design specification laying down the modifications and improvements made to previous models.

Physically, the Type 79 was very similar to its predecessors, and can be difficult to distinguish. While still a first-generation MBT, it was a huge improvement on earlier Chinese tanks, with stabilization, optics and improved fire control allowing the crew to take advantage of the 105mm (4.1in) rifled gun's capabilities.

The Type 79 benefited from a more powerful engine, delivering 544kW (730hp) rather than the 432kW (580hp) of the Type 69-II model. According to Chinese sources, road performance was much the same, but additional power presumably improved cross-country performance. Other developments included the provision to fit ERA and slight modifications to the turret, which may have allowed additional electronic systems to be fitted. Night-vision equipment for the commander, driver and gunner enabled the Type 79 to move and fight effectively at night.

Defensively, the Type 79 retained the capability of creating a smokescreen by injecting fuel into the exhausts, and mounted smoke grenade dischargers that had first appeared on earlier tanks in its family. Removable side skirts offered some protection from shaped-charge weapons, as did the equipment racks along the sides of the turret.

Missile tank

The Type 79 was introduced at a time when guided missiles were emerging as an extremely serious threat to armoured vehicles. The first tanks in its family were developed at a time when high-velocity projectiles from tanks or anti-tank guns were the main hazard, with the relatively small warheads of infantry weapons often defeated by thick armour. The Arab–Israeli war of 1973 demonstrated how devastating even the crude guided missiles of

Type 79

This late-production Type 79 with a three-tone camouflage dates from the 1990s. Although sometimes considered a separate model, the Type 79 might better be thought of as the final result of all the Type 69 upgrade programmes that had gone before.

Type 79
Date 1984
Number Produced Not Known
Weight 37.2 tonnes (36.7 tons)
Length (with gun) 8.58m (28ft 2in)
Width 3.31 (10ft 10in)
Height 2.8m (9ft 2in)
Engine Type 79 diesel 544kW (730hp)
Maximum Road Speed 50km/h (28mph)
Crew 4
Armour Up to 180mm (7in)
Main Armament Type 83 105mm (4.1in) rifled gun
Main Gun Ammunition Stowed Not Known

the time could be, with increased chances of being hit even on the move compounded by larger warheads than a shoulder-launched weapon could carry. Sophisticated reactive armour could disrupt the plasma jet generated by the missile's warhead, while one that struck a skirt or the crew's belongings would focus on a point short of the main armour protection.

This was a world where infantry posed an increasing threat to armoured vehicles, and in which some Soviet experts described anti-tank equipment as 'the most important weapon'. The Type 79 had to be able to survive in this environment; in many ways, its design features reflect the changes in the combat environment that had occurred since its ancestor, the T-54, was developed towards the end of World War II.

Most sources state that all Type 79 MBTs were retained by Chinese forces. Type 79s are also listed as serving with Pakistan, but these were improved Type 69s with more or less the same specification and not 'officially' Type 79s. Although an improvement on previous designs, the Type 79 was something of a transition model and was quickly supplanted in Chinese service by more advanced second-generation tanks.

Type 80

The Type 80 was developed in response to a need for a new generation of main battle tanks that would incorporate lessons learned with previous designs and that could cope with the changing battlefield environment of the 1980s.

Although the Type 59/69/79 family was used as a starting point for the design, the Type 80 was a new tank intended from the outset to incorporate features that had been added to earlier designs wherever they could be made to fit.

Hull form resembled the Type 69, retaining the low silhouette and rounded turret of its predecessors, but had an all-new hull design using welded construction. A larger powerpack, derived from a German diesel engine, necessitated a slight lengthening of the hull. The use of six smaller roadwheels compared to the five larger ones of the previous classes altered the tank's appearance to some degree. A new torsion bar suspension system was developed, improving cross-country performance, and with a more powerful engine than the Type 79 and only slightly increased overall weight, top speed was also improved to around 55km/h (35mph).

Prototype

The Type 80 reached the prototype stage in 1985 and was finalized three years later, making it a slightly newer contemporary of the US M1 Abrams MBT. The Type 80 was less technically ambitious, retaining a diesel powerplant rather than the gas turbine of the Abrams, and overall representing a more gradual approach to improvement than the risky and expensive leaps taken by some other nations.

Whereas the Abrams and other Western tanks of the era were built with composite armour as standard, the Type 80 had the capability to fit it but used more conventional protection as standard. Composite armour is difficult to produce, and can be best cast in

Type 80
Date 1981
Number Produced 1800
Weight 38.6 tonnes (38 tons)
Length (with gun) 9.32m (30ft 7in)
Width 3.37m (11ft 1in)
Height 2.29m (7ft 6in)
Engine 12150L-7BV diesel, 544kW (730hp)
Maximum Road Speed 55km/h (35mph)
Crew 4
Armour RHA + side skirts
Main Armament 105mm (4.1in) Type 83 rifled cannon
Main Gun Ammunition Stowed 44

Type 80

An early Type 80 MBT, without side skirts to show the drive-train. There is a gap between the second roadwheels and the four rear ones.

Type 80

A Type 80 kitted out for a parade in the early 1980s.

Type 80

A Type 80 with the standard three-tone camouflage (dark sand, olive green, dark grey). The new generation of Chinese tanks were readily identifiable from their predecessors by the use of six smaller road wheels instead of five large ones.

large, flat slabs that are not compatible with the idea of a rounded turret to deflect incoming projectiles. Its near-universal adoption gives many modern main battle tanks a generally similar appearance, making the Type 80 readily identifiable as a product of an earlier generation.

The tracks of any tank are a potential weak point. During World War II, seemingly invincible Soviet and German heavy tanks were disabled by shooting at the tracks and then bypassed or eliminated by infantry using satchel charges. This danger has never passed, and in today's environment infantry have many weapons capable of damaging tracks or even penetrating the sides of the hull. The danger of close-range infantry attack grows ever greater as tanks are increasingly forced to operate in an urban environment.

Armour

Many armoured vehicles today use 'bar' or 'slat' armour to defeat rocket-propelled grenades; racks around the sides of the turret provided similar protection as well as stowage for large pieces of equipment and the possessions of the crew. Meanwhile, protection against shaped-charge weapons and explosives that might be thrown onto the tracks by enterprising infantrymen was provided by skirts, which were hinged to allow access for maintenance.

For its main armament, the Type 80 used the same 105mm (4.1in) rifled gun as its predecessor. Derived from the Austrian version of a proven NATO weapon, this gun was capable of successfully engaging any armoured vehicle of the time with kinetic energy penetrator ammunition or shaped-charge HEAT rounds. Explosive and fragmentation rounds were available for softer targets, but in general most of the tank's 44 carried rounds would be for anti-tank combat.

Type 80-II Prototype

The new roadwheel arrangement was more difficult to produce but resulted in a smoother ride and therefore better accuracy when firing on the move as well as less crew fatigue.

Type 80-II

The Type 80-II was an experimental model incorporating a variety of upgrades. It was not adopted for service but provided a testbed for equipment used elsewhere.

A Type 80 MBT undergoes testing. The Type 80 was a capable design offering many advantages over earlier Chinese MBTs. However, it was still less sophisticated than Russian and Western tanks of the same era.

A gun is only as good as its fire control system allows, and that of the Type 80 was a step forward from its predecessors. Both the weapon and the gunner's sight were stabilized, improving the chances of a first-shot hit even on the move. On the Type 80, like its predecessors, the gun was manually loaded, requiring a crew of three – commander, gunner and loader – in the turret. As a secondary responsibility, the loader also manned the 12.7mm (0.5in) heavy machine gun when short-range air defence proved necessary.

Service history

Few Type 80s were exported, but around 3000 were constructed before production moved on to other designs. The development of the tank reflected changes in the world political climate; the Type 80-II model incorporated a laser rangefinder licensed from the UK. Like many modifications, this was something of a compromise. Although it improved combat effectiveness it had to be externally mounted, which left it vulnerable not only to combat damage but also to environmental hazards.

Development of the Type 80-II model was an opportunity to experiment with modifications that would become standard on following designs. Sources vary, but it seems that the Type 80-II saw some rearrangement of armour and alterations to the transmission and suspension along with minor improvements in protection for critical but vulnerable components such as optics.

At the time the Type 80 entered service, China was still lagging behind the West in terms of technical sophistication and experience of integrating new systems. It represented a very significant improvement over the preceding classes, but still belonged to an earlier generation than, say, the M1 Abrams. It is notable that although the Abrams was repeatedly updated and remains a first-class fighting platform to this day, the newer Type 80 was quickly overtaken by advancing technology and sidelined. However, the Type 80 did serve as a development platform for the following Type 88 and Type 85 main battle tanks, and was a testbed for features that would become standard on third-generation Chinese tanks. As such, it was an important transition model even though it never saw action in Chinese hands.

Type 80B

China has a long tradition of showing off its best military hardware at its National Day parade, not coincidentally allowing foreign observers to be impressed with advanced vehicles like this Type 80B.

Type 88

Entering service in 1988, the Type 88 MBT was an improved version of the Type 80. The two are often considered to be versions of the same vehicle, and usually referred to as Type 80/88.

Visually, the Type 88 was almost identical to the Type 80, with the same general lines and six large roadwheels. The latter made it relatively easy to distinguish from members of the Type 59 family. Where the Type 80 was obviously of Soviet descent, the new Chinese tanks had many Western features.

Laser rangefinder

Development from the Type 80 allowed the laser rangefinder to be incorporated into the turret and thus much better protected, while the turret itself gained ERA at the expense of some of the stowage racks. The biggest change was inside the turret, where an autoloader allowed the loader's position to be deleted. The turret-roof heavy machine gun, previously operated by the loader, became the commander's responsibility.

Comparison with the contemporary Challenger I, which entered British service in 1982, is inevitable. The Challenger featured all-round composite armour, a 120mm (4.7in) main gun and hydropneumatic suspension, whereas the Type 88 used torsion bars.

Both were the most advanced tanks fielded by their parent nations, but whereas the Type 88 did not exceed 40.6 tonnes (40 tons) in any of its variants – only a little more than preceding Chinese MBTs – the Challenger I had grown to 63 tonnes (62 tons) by way of its 56-tonne (55-ton) predecessor the Chieftain.

Light tank

How much a part the Chinese terrain played in the decision to keep weight down to this modest level is open to speculation. Technical limitations, notably in terms of engine power, suspension and power-to-weight ratio, may have also been a factor, but even once very powerful engines were available Chinese tanks remained relatively light compared to Western designs. The Soviet T-80, dating from around the same period, weighed in at 42.6–46.7 tonnes (42–46 tons). In this regard, the Type 88 remained closer to the Soviet design philosophy than a Western one.

Type similarities

Some sources state that the Type 88 was the version adopted by the Chinese military, and the Type 80 was an export version, while others place both in Chinese service with few exported of either model. The line between the Type 80 and Type 88 may have been blurred by upgrade programmes or the difficulty of distinguishing between similar vehicles operated by a military that is unforthcoming with detailed information.

In any case, development did not stop with the Type 88. Further enhancements included a longer gun with a higher muzzle velocity, and a redesigned breech system that improved the autoloader's performance. This increased effective rate of fire from eight to 13 rounds per minute. These upgrades produced the Type 88A and B models; their similarities make it difficult to determine which are in service in any given area, although it is known that Myanmar took more than 200 Type 88B tanks. These may have seen active service during their career.

Confusingly, perhaps, the Type 88C model was derived from the Type 85 MBT. It was generally similar to the Type 88 but armed with a 125mm (4.9in) smoothbore gun, and later gained a more powerful engine that was retrofitted to existing tanks of this model.

Type 88

Date 1988
Number Produced 1860
Weight 39.6 tonnes (39 tons)
Length (with gun) 9.32m (30ft 7in)
Width 3.37m (11ft 1in)
Height 2.29m (7ft 6in)
Engine VR36 V-12-cylinder diesel, 544kW (730hp)
Maximum Road Speed 55km/h (35mph)
Crew 4
Armour Min–max RHA 20–203mm (0.8–8in) + composite + ERA
Main Armament 105mm (4.1in) Type 83 rifled cannon
Main Gun Ammunition Stowed 48

Type 88

The Type 88 was a development of the Type 80, representing the typical Chinese approach of steady and reliable improvement rather than revolutionary leaps forward.

Type 88A

The Type 88A gained a new, longer 105mm (4.1in) gun with an improved thermal sleeve. Both factors improved accuracy when firing at long ranges.

Type 85

The Type 85 main battle tank might be thought of as a distant cousin of the Type 80 and its derivatives.

A first prototype, designated Type 85-I, was tested and found wanting, but the Type 85-II was deemed successful and entered service in 1991. The requirements placed upon this new tank design resulted from concerns about the ability of Type 80/88 MBTs to survive against and successfully engage the latest generation of tanks emerging worldwide. Although certainly an improvement on earlier vehicles, the Type 80/88 still lagged behind its contemporaries. It appears that the Type 85 was an interim model intended to pave the way for an entire new generation of Chinese tanks, integrating systems trialled in the Type 80/88 into a new and far more effective package.

The 1991 Gulf War permitted Chinese designers to make the most direct comparisons possible between MBTs operated by the Coalition powers and Iraqi vehicles that were mostly Soviet and Chinese in origin. The ability of modern tanks to engage and destroy superior numbers of older designs was graphically demonstrated; in particular, the traditional Soviet-style cast rounded turret was found to offer little protection.

The vehicle that entered production integrated proven systems into a new package. Its hull, redesigned but retaining the general lines and dimensions of the preceding classes, was built around the engine and drive train of the Type 80/88. Overall weight also remained about the same, increasing only slightly to 39.6 to 40.6 tonnes (39–40 tons). The main and obvious difference between the Type 85 and preceding designs was an entirely new turret of angular appearance rather than the previous rounded design.

Composite armour

Whereas preceding tanks had made some use of composite armour, the Type 85 was designed from the outset to use it. Hull protection was improved by the use of composite armour, usually with a layer of reactive armour on top. The cast, rounded turret was finally replaced with a welded version, with particularly thick armour on the front faces. Some sources suggest that the armour specification included the ability to defeat a 125mm (4.9in) discarding sabot round striking the turret face.

Initially, the turret held the same 105mm (4.1in) gun that had armed preceding designs, but tests and experience gained elsewhere in the world indicated that this weapon was not adequate to penetrate the best-protected modern tanks. A 125mm

Type 85
Date 1989
Number Produced 600
Weight 39.6 tonnes (39 tons)
Length (with gun) 9.33m (30ft 7in)
Width 3.37m (11ft)
Height 2.29m (7ft 6in)
Engine 12150L-7BV diesel 544kW (730hp)
Maximum Road Speed 55km/h (35mph)
Crew 4
Armour Not Known
Main Armament 105mm (4.1in) Type 83 rifled cannon
Main Gun Ammunition Stowed 46 rounds

Type 85 Prototype
Concerns about the vulnerability of the Type 80/88 led to the more survivable Type 85 starting development before its predecessor had entered service.

Type 85-I

The Type 85-I gained a thermal sleeve for the gun and a modified turret incorporating composite armour.

(4.9in) main weapon was substituted, and was given an autoloader, which allowed the loader's position to be deleted. It is probable that this gun and loading mechanism were derived from those used on the Soviet T-72.

The 544kW (730hp) diesel engine of the Type 80/88 was retained, along with a similar arrangement of roadwheels. The slightly lengthened hull required some minor adjustments in suspension, but overall the proven system of the preceding classes was retained. Similarly, most of the tank's systems had been trialled in the Type 80/88 and were well proven, but were integrated to create a more effective and better-protected package. Fire control was an improvement over preceding designs, using electronic systems of Western origin. The end result was a tank of characteristically Chinese design, combining a Soviet-style gun directed by Western electronics and a European powerplant working through a drive train that could trace its pedigree back to Soviet origins. All this was put together to create a vehicle that was tailored to Chinese needs, and that would serve as a stepping stone to the next generation of main battle tanks.

NBC protection

During the Cold War, it seemed likely that armoured fighting vehicles would have to operate in a heavily contaminated environment. Indeed, many observers grimly stated that all major future conflicts would be nuclear. This, thankfully, has not happened, although chemical weapons have been used in various conflicts and undoubtedly will continue to do so despite being banned by international treaties.

Early NBC protection for tank crews was individual, requiring crewmembers to wear bulky equipment while operating in a very tight space. Collective protection,

Type 85-II

Upgrades to the Type 85-II were mainly electronic. It gained improved fire control by way of a ballistic computer and supporting sensors.

created by sealing the tank and creating overpressure inside to prevent agents entering, allows a much more comfortable and therefore efficient operating environment, although decontamination is required after passing through an affected area. An effective, integrated NBC protection system is necessary to any modern tank design, and that of the Type 85 reflected years of experience at integrating such peripheral but vital systems into a vehicle without interfering with its main function.

New capability

The Type 85 created an effective modern battle tank that was no longer tied to a Soviet 'tank swarm' style of combat. Increased depression enabled firing from a hull-down position behind a ridge or slope, and if the Type 85 did have to shoot on the move, its chances of a hit were greatly improved. Combined with a hard-hitting main gun, this meant that Chinese tanks were individually far more capable than they had been in the past. Fielding large numbers remained an option, of course, but even if logistics or availability limited the size of the force that could be deployed it would remain highly effective.

Some Type 80/88 tanks were upgraded to Type 85 standards, and overall the design proved sufficiently satisfactory to achieve some export success. Pakistan was the main client, manufacturing a licence-built version designated Type 85-IIAP, while Sudan

The Type 85-IIAP is a license-built version produced in Pakistan. Like the Chinese Type 85-IIM, it is armed with a 125mm (4.9in) smoothbore gun.

took delivery of a quantity of Type 85s direct from China. These saw action in the Sudanese Civil War.

Type 85-III

Development of the Type 85 did not stop, with a Type 85-III appearing in 1995. It featured a 746kW (1000hp) engine, which initially caused some troubles, and may have been aimed at the export market. This specification was part of the development process that led to the Type 90 and Type 96 main battle tanks, and became the standard to which many Type 85s were upgraded.

The close family relationship between many Chinese tanks, and the tendency of older examples to be upgraded to newer specifications, creates confusion about production numbers or the exact models in service at any given time and place. Chinese sources are not forthcoming about specifications or numbers, so published figures on numbers or even models tend to vary considerably.

Often, sources cannot agree on which model of a given tank design was exported to a particular client, and production numbers remain equally vague. For example, estimates of the total production of Type 85s vary from a few hundred to more than 2000, and in-service numbers may include upgraded examples from the Type 80 or Type 88 family.

Type 85-III
Date 1989
Number Produced 600 (all types)
Weight 39.6 tonnes (39 tons)
Length (with gun) 9.33m (30ft 7in)
Width 3.37m (11ft)
Height 2.29m (7ft 6in)
Engine 744kW (1000hp) diesel
Maximum Road Speed 55km/h 35mph
Crew 4
Armour RHA classified, BAR, ERA, rubber side skirts
Main Armament Type 83 105mm (4.1in) rifled gun
Main Gun Ammunition Stowed 46 rounds

Type 85-III Prototype
The Type 85-III had a difficult development process which eventually established it as an upgrade specification for the Type 85 family.

Type 85-IIM ERA
The Type 85-IIM carries a 125mm (4.9in) gun and has the capability to mount additional explosive reactive armour (ERA) in easily replaceable blocks.

Type 90

The Type 90 is the first of the current generation of Chinese main battle tanks. It represents the coming of age of Chinese tank design, leaving behind holdovers from the original Soviet-derived designs.

This process was not abrupt. The Type 90 was developed from a version of the Type 85 designated Type 85-IIM and incorporates several systems used in earlier models.

Extensive use is made of composite armour. This is applied over the top of standard steel armour in modules that can be removed for maintenance or to lighten the tank for transportation. Damaged modules can also be swapped out or replaced with armour more suited to the tank's current role. With a full set of composite armour modules in place, the Type 90 weighs 48.7 tonnes (48 tons), significantly more than previous Chinese designs but still much less than most Western tanks.

The Type 90 mounts the same 125mm (4.9in) smoothbore gun used on the Type 85, and can deliver a range of munitions. Any main battle tank must be able to fight other tanks, but it requires additional capabilities to justify its huge cost. Thus the Type 90 can fire armour-piercing fin-stabilized discarding sabot (APFSDS) and HEAT ammunition against armoured targets, and has a high explosive fragmentation (HE-frag) round available for other purposes. In addition to challenging other tanks for dominance of the battlefield, the Type 90 is intended to be able to support infantry by tackling bunkers and strongpoints.

The 125mm (4.9in) gun can also launch guided missiles. Opinions are divided about such weapons, and they have not had a trouble-free development. However, in theory at least, a gun-launched missile provides additional capabilities not available with conventional projectiles. A tank equipped with such weapons may be able to shoot down attack helicopters – a highly desirable capability in an environment where tank-hunting gunships are an ever-present threat. The Type 90 uses a locally produced version of the Russian 9M119 *Svir* (or *Refleks*) missile, a laser-guided weapon with a range of 4–5km (2.5–3 miles).

Export model

The original Type 90 was not adopted for service with the Chinese armed forces, probably due to issues with the powerplant that may have affected

Type 90-I Prototype

The Type 90 was developed from the Type 85, gaining modular composite armour for the turret front and additional reactive armour on the hull. This prototype was solely intended for export to Pakistan in 1997.

mobility. Several different engines and associated equipment were tried, including powerplants used on the Challenger and Leclerc MBTs. Eventually an 895kW (1200hp) diesel plant was installed, giving a 65km/h (40mph) top speed and rough-terrain performance equivalent to earlier designs.

Beyond a certain point, a tank's ability to deal with obstacles is limited by its suspension, ground clearance and hull size rather than power-to-weight ratio, and the powerplant's other characteristics become increasingly important. The Type 90 was designed with ease of maintenance in mind, increasing serviceability rates. Fuel economy represents half of the operating radius equation, along with the amount of fuel the tank can carry. The Type 90's radius of approximately 500km (311 miles) is equivalent to most other modern armoured fighting vehicles and compares favourably to previous generations of Chinese MBTs.

MBT-2000

Although the Type 90 was not taken into Chinese service, the design was further developed with a view to export sales. This programme was a success, although at times it fell foul of international politics. The Type 90-IIA model, developed to suit Pakistani requirements, featured a 1119kW (1500hp) diesel plant of French origin.

However, it could not be delivered due to French sanctions resulting from Pakistan's nuclear programme. The Type 90-IIM was then developed using a Chinese-built engine. It was offered to Pakistan under the project title MBT-2000, but reliability issues delayed the project still further.

Finally, a new model was adopted for service with Pakistan's military under the designation Al-Khalid. The MBT-2000 also found favour with Morocco and Bangladesh.

As the Al-Khalid, the MBT-2000's most likely opponents are Indian T-72s and T-90s, or the Arjun, which may be the world's heaviest tank. This situation reflects the greatly changed world in which modern tanks must operate. In the heavily polarized environment of the Cold War, it was possible to predict what sort of equipment a potential enemy would be using based on its alignment with the Communist or Western powers. This in turn allowed a design to be tailored to a particular doctrine and to counter specific threats. That likelihood no longer

Al-Khalid tanks participate in a Pakistani national day parade, 2017.

Type 90-II

The Type 90-II suffered from technical problems arising from the difficulty of developing a sufficiently powerful engine that could function in all the conditions the tank was likely to encounter.

exists, so the modern MBT must be able to deal with a wide range of potential opponents – especially if it is intended for the export market.

Export success

The Al-Khalid/MBT-2000 has achieved some additional export success, having been adopted by Bangladesh, Morocco and Myanmar. The basic design continues to be developed, by way of the Al-Khalid I, which features an advanced autoloader and improved fire control. The Al-Khalid II benefits from work done on the Type 99 MBT and uses some components developed for it. Earlier MBT-2000s will likely be upgraded to this standard, although they may or may not receive the new 1119kW (1500hp) engine.

Type 90

Number Produced Not Known
Weight 48.7 tonnes (48 tons)
Length (with gun) 10.28m (33ft 8.5in)
Width 3.45m (11ft 4in)
Height 2.3m (7.5ft)
Engine 6TD diesel 900kW (1200hp)
Maximum Road Speed 65km/h (40mph)
Crew 3
Armour Not Known
Main Armament 120mm (4.7in) smoothbore cannon
Main Gun Ammunition Stowed 39 rounds

Type 90-IIM

Engine problems were tackled for the Type 90-IIM model by using a powerplant of Ukrainian design. It was adopted by Pakistan as the Al-Khalid.

Type 96

The Type 96 MBT was developed from the Type 85, though with influences from other projects.

It has become the mainstay of Chinese armoured forces, with estimates ranging from 1500 to more than 3000 vehicles produced. This makes the Type 96 the most important Asian tank currently in large-scale service.

In terms of layout and design, the Type 96 is not innovative. It uses a standard MBT layout with the driver seated at front left in the hull and the commander and gunner in the turret.

Type 96

The Type 96 is armed with a 125mm (4.9in) smoothbore cannon supported by an autoloader derived from that of the Russian T-72.

The 125mm (4.9in) smoothbore gun is fed by an autoloader and can probably launch missiles in addition to firing a range of ammunition types. The gun and autoloader are based on systems derived from the Soviet T-72, though local development continues and systems tend to diverge from their origins over time.

The nature of tank evolution has changed since World War II. In the early to mid-twentieth century, development was typically characterized by the addition or alteration of heavy or bulky systems. In many cases, the upgradability of a tank design was

Type 96
Date 1997
Number Produced 1200
Weight 43 tonnes (42.2 tons)
Length (with gun) 10.28m (33ft 9in)
Width 3.45m (11ft 4in)
Height 2.3m (7ft 6in)
Engine Norinco 12150ZL V12-cylinder turbo-charged diesel, 582kW (780hp)
Maximum Road Speed 65km/h (40mph)
Crew 3
Armour Modular RHA + ERA + side skirts
Main Armament 125mm (4.9in) Type 90 L/48 smoothbore cannon
Main Gun Ammunition Stowed 45

Type 96 (ZTZ-96)

The Type 96 came to occupy a niche below that of the elite armoured forces, equipping some units while others made do with older designs, including the venerable Type 59 family.

limited by its ability to receive a larger gun. A small turret ring limited the size of the turret, which in turn made it impossible to fit a gun needing more than a short recoil space. Advances in armour protection on enemy tanks quickly made such vehicles obsolete, at least in their intended role.

Gun sizes in particular increased enormously during this period. A 37mm (1.6in) main gun was considered acceptable in 1940; in 1945, calibres around 75mm (2.9in) were common, and the first-generation main battle tanks appearing in the early Cold War years were carrying guns in the 90–105mm (3.5–4.1in) class. The leap to 120mm or 125mm (4.7in or 4.9in) guns established a certain minimum size of turret for an MBT, but despite the occasional experiment this has remained stable ever since.

The size of turret ring required to accommodate these weapons to a great extent dictates the overall dimensions of the tank, which in turn has implications for its final weight. Although some compromises can be made regarding armouring, to be survivable on the modern battlefield a tank needs a minimum level of protection, which results in the requirement for a large powerplant and heavy suspension. The firepower/protection/mobility trade-off is nothing new, but in the modern environment the amount of variation among answers is very limited.

A Type 96 competing in the Tank Biathlon at the International Army Games in 2016. The Chinese team took second place.

Technical improvements

Most improvements in modern tanks tend to be electronic or technical rather than physical. Improved fire control, battle management or vision systems can make a vehicle far more effective without significantly altering its physical characteristics. Similarly, ease of maintenance may or may not make a difference to the effectiveness of a single tank, but the armoured force as a whole will be greatly enhanced by improved serviceability rates. It is not always obvious to the casual observer when one armoured fighting vehicle is

Type 96A1

Date 2007
Number Produced 326
Weight 44.5 tonnes (43.8 tons)
Length (with gun) 10.28m (33ft 9in)
Width 3.45m (11ft 4in)
Height 2.3m (7ft 6in)
Engine Norinco 12150ZL V12-cylinder turbo-charged diesel, 582kW (780hp)
Maximum Road Speed 61km/h (38mph)
Crew 3
Armour Modular RHA + ERA + ECM + side skirts
Main Armament 125mm (4.9in) Type 90 L/48 smoothbore cannon
Main Gun Ammunition Stowed 44

superior to another in this regard, but for the operator serviceability rates, availability of spares, ease of upgrade regarding electronic systems and overall lifespan expectations can be of great importance.

The Type 96 was developed with these factors in mind. Its composite armour can be removed partially or wholly, and replaced or upgraded as new composites become available. Similarly, the powerplant and transmission were designed as a single unit with ease of maintenance in mind. Chinese sources state that the entire unit can be swapped out in around 40 minutes at a field workshop. The 746kW (1000hp) diesel engine was developed on preceding designs and is a mature system, although Type 96s exported to Sudan appear to have been given a 544kW (730hp) powerplant instead.

Good sensors and fire control, backing up a powerful main gun, can permit a tank to protect itself by breaking the kill chain – spotting the enemy first and disabling their vehicles. However, there can be no reasonable expectation that this will always be achieved, and it must be assumed that a tank will come under attack at some point. Defence is thus often a matter of defeating a warhead or preventing penetration, but it is vastly preferable not to be hit at all.

Avoiding being hit has traditionally been achieved by concealment or

Type 96A
The Type 96A is a second-batch production model incorporating improved modular armour and a more powerful engine. It was still considered underpowered by some observers.

Type 96A1
The Type 96A is a second-batch production model incorporating improved modular armour and a more powerful engine. It was still considered underpowered by some observers.

Type 96G

Production of the Type 96 ended in the early 2000s, but was recommenced with a new Type 96G model which was adopted by many first-echelon armoured units.

speed of movement. Camouflage can only do so much, especially in an age of advanced sensors, and even with the thermal-masking technologies currently making their way onto the battlefield a tank can expect to be spotted sooner or later. Movement might throw off an enemy's aim, but modern fire control systems are easily able to compensate. In the case of guided weapons, a first-shot hit is very likely, and as the world's electronics industries advance, even relatively unsophisticated militaries are able to field at least basic guided weapons.

Smoke can defeat laser-guidance systems and will increase the targeting difficulties facing enemy gunners. Smoke grenade dischargers are standard equipment on most tanks, and many designs can create smoke using their own hot exhaust system. However, new electronic options are becoming available. The Type 96 is fitted with devices designed to interfere with the guidance systems of missiles and to disrupt laser designation.

Upgrade programme

This enhanced defensive capability was implemented as part of the upgrade programme that produced the Type 96A and Type 96G models. In addition to improved armour protection on

Type 96G
Date 2007
Number Produced 487
Weight 44.6 tonnes (43.9 tons)
Length (with gun) 10.28m (33ft 9in)
Width 3.45m (11ft 4in)
Height 2.3m (7ft 6in)
Engine Norinco 12150ZLD V12-cylinder turbo-charged diesel, 746kW (1000hp)
Maximum Road Speed 66km/h (41mph)
Crew 3
Armour Modular RHA + ERA + side skirts
Main Armament 125mm (4.9in) Type 90 L/48 smoothbore cannon
Main Gun Ammunition Stowed 44

VT-1

Although possibly underpowered, the Type 96 (and the VT-1A export model pictured here) is overall an effective AFV and has been adopted by the Royal Moroccan Army.

the turret, the latest version of the Type 96 gained electronic defensive systems, a semi-automatic combat management system and improved vision and sighting devices. However, increased weight led to a reduction in performance that, although not serious, has led to the Type 96 being considered underpowered by some observers.

The Type 96 is primarily used by the Chinese armed forces, where it has replaced older tanks in most formations. The only export customer to date is Sudan, where the tank has been tested in combat. Although a small-scale conflict can provide only limited data, the Type 96 seems to have proven greatly superior to the T-72s it encountered. This is despite the fact that Type 96s in Sudanese service have much less powerful engines than the Chinese version and are thus likely to suffer mobility issues.

The sale of MBT-2000 variants to overseas users has created some confusion about exactly which nations are using what model. Although the MBT-2000 is in theory a different tank, upgrade packages bring some variants very close to the Type 96 in terms of equipment and performance. If capabilities and characteristics are considered rather than the original specification to which a tank was built, the lines between various Chinese MBT designs become very blurry indeed.

During the 2016 Army Games at the Alabino Firing Range, Russia, China's team celebrate their second place ranking on the turret of a Type 96A MBT.

Type 99

Any nation operating tanks would prefer that they were all of the same model, and for that model to be the most advanced possible. However, the expense of maintaining and regularly replacing a tank fleet makes this impractical.

Older tanks can be kept in service for many years and, with relatively inexpensive upgrades, can remain quite effective against most opponents. Even when overmatched, older tanks can serve their nation well by providing at least a modicum of armoured capability and backing up the spearhead units.

Thus many nations have moved, deliberately or otherwise, to a 'two-tank' or high-end/low-end model. For a minor nation with a very limited budget this can mean a cycle of obtaining a few fairly modern tanks – often obsolescent and overmatched in the superpower arena but potent compared to vehicles used by other powers in the area – and patching up older vehicles until they become unserviceable.

There is a point beyond which the cost of keeping elderly vehicles in service, let alone effective, is prohibitive. China's solution to this problem is to use the Type 96 for most of its armoured forces and to create a more advanced and capable armoured fighting vehicle, the Type 99, for its elite formations. Both have the potential to remain in service for a long time until eventually they will need to be sold off and a new generation of armoured vehicles will take their place.

Type 98 prototype

The Type 99 was developed by way of an interim design designated Type 98, which first appeared in public in 1999. Although shown off to the world alongside in-service vehicles, the Type 98 was not adopted by the PLA. Instead, only a small batch – probably numbering about 18 tanks – was produced for development and evaluation purposes.

The Type 99 followed in 2001 and remains in service with the most prestigious of China's armoured units. Compared to the Type 98, it features a far more powerful engine, delivering 1119kW (1500hp) rather than the previous model's 895kW (1200hp). This increases maximum speed from around 65 to 80km/h (40 to 50mph), and improves cross-country performance accordingly. The new engine is based upon German

Type 98
Date 1999
Number Produced 42
Weight 53.9 tonnes (53 tons)
Length (with gun) 11m (36ft)
Width 3.5m (11ft 6in)
Height 2.25m (7ft 3in)
Engine Norinco 12150ZL V12-cylinder turbo-charged diesel, 1119kW (1500hp)
Maximum Road Speed 80km/h (50mph)
Crew 3
Armour Modular composite + modular spaced + ERA modules + side skirts
Main Armament 125mm (4.9in) Type 99 L/80 smoothbore cannon
Main Gun Ammunition Stowed 41

Type 98 Prototype
In a move typical of Chinese tank design philosophy, the Type 98 was developed as an interim model and used as the basis of the Type 99.

Type 99

The Type 99 can be distinguished from its predecessor by the turret front, which is shaped to give a greater chance of deflecting kinetic-energy penetrators.

technology, and the Type 99 overall is influenced by the latest incarnation of the German Leopard 2A5 MBT. This is obvious from the arrangement of the ERA blocks on the turret. Protection goes beyond solid and reactive armour, however.

The Type 99 has laser-warning devices fitted, which are becoming increasingly common on combat vehicles. Whether it is a gun's rangefinder or a designator guiding a missile or bomb, a laser striking a tank is a clear warning that defensive measures need to be taken.

In addition to the traditional methods of dodging behind an obstruction or laying smoke to break the lock, the Type 99 is equipped with laser dazzler devices. These are intended to 'blind' sensors by overloading them. The concept is not new; British warships attempted to dazzle Argentinian pilots during the 1982 Falklands War, and various attempts have been made to use lasers as less-lethal weapons since. In many cases, these efforts were made on an ad-hoc basis using equipment not designed for the purpose, and could fall foul of treaties banning the use of lasers capable of causing permanent blindness as weapons.

Perhaps more importantly, manually aimed lasers are not capable of reacting quickly and accurately enough to defeat an electronic sensor and associated guidance system. Modern laser dazzlers are able to do so, and are becoming increasingly important defensive systems as the amount of laser-guided weaponry available to even minor powers continues to rise.

Increased integration of the tank's sensor systems allows the commander to use the gunner's sight and thus see for himself exactly what the weapon is doing. Fire control uses a target tracking system that may be capable of engaging in hunter/killer operations. In this mode, the gun can be instructed to automatically engage targets of a certain type when they are recognized, with priorities or other considerations programmed by the gunner. This permits extremely fast responses to targets of opportunity or a rapidly changing situation such as an ambush.

Upgrades

Systems integration of this sort greatly enhances the combat effectiveness of armoured fighting vehicles, and while expensive to develop adds little

Type 99

Date 2001
Number Produced 500
Weight 54 tonnes (53.1 tons)
Length (with gun) 11m (36ft)
Width 3.5m (11ft 6in)
Height 2.25m (7ft 3in)
Engine Norinco 12150ZL V12-cylinder turbo-charged diesel, 1119kW (1500hp)
Maximum Road Speed 80km/h (50mph)
Crew 3
Armour Modular composite + modular spaced + ERA modules + side skirts
Main Armament 125mm (4.9in) Type 99 L/48 smoothbore cannon
Main Gun Ammunition Stowed 42

weight to the tank. Upgrades may take the form of software or very small electronic systems, and may represent only a slight increase in capability for a very large research investment. However, this is the nature of modern tank development.

The huge increases in hitting power that went with a doubling of gun calibre, or the defensive advantages gained from sloped armour and the use of composite materials, required large increases in size and weight, a process that seems now to have run its course. Unless a quantum leap in capability occurs due to some new weapon or materials technology, further development is likely to be costly in time and money rather than weight and bulk, and will be incremental rather than representing a great leap forward in capability.

However, the combination of small improvements in ammunition, sensors

Type 99KM

The Type 99 KM features upgraded armour and electronics. It has been the subject of greatly exaggerated claims about its performance. This example sports a distinctive 'crude-digital' four-tone camouflage scheme of three shades of green combined with sand yellow.

Below: Type 99A main battle tanks took pride of place at parades held in 2015, celebrating the 70th anniversary of the end of World War II.

and battle management technology can add up to a huge advantage over potential opponents – and vice versa. Thus the process must continue, and huge budgets are eaten up to create often intangible results. It is only when a tank design is committed to battle that this investment finally shows its worth.

The Type 99 also makes use of lasers for communication. Mounted on the rear of the turret is a laser communication system. Although limited to line-of-sight, this system is virtually impossible to intercept or even detect. It can be used passively for identification purposes, which goes beyond the admittedly important function of preventing friendly fire. The battle management system in each tank knows exactly where others in the same formation are located, enabling the commander to make decisions about how to manoeuvre his own tank, or a senior officer to decide where to put elements of the force in order to obtain advantages over an opponent.

Support assets such as airstrikes and artillery fire can be directed with great precision using this system; the combination of highly accurate weaponry and certainty about the identity and location of friendly forces allows supporting fire to be called in on targets far closer than would have previously been possible, which has defensive advantages as well as shortening the time between 'softening up' with artillery or air strikes and an armoured assault. An enemy force still emerging from cover or dazed by bombardment will not defend as effectively as one that has returned to its damaged positions.

Type 99A and Type 99G

The Type 99 was a highly capable armoured fighting vehicle when it entered service in 2001, but naturally development did not stop. In 2007–8 a new version, variously designated Type 99A and Type 99G, began to emerge. It is believed that this version included a separate thermal sensor for the commander and electronic upgrades. The Type 99A1 that followed featured an altered turret design and possible electronics upgrades, and in 2010 the next upgrade version entered trials. This vehicle is variously designated Type 99A2 or Type 99KM.

VT-4

The VT-4, originally designated MBT-3000, was developed for export by way of the VT-1A. These tanks

VT-4

The VT-4 is an export model derived from the MBT-2000 variant of the Type 90 family. It has only recently entered the market and has yet to prove itself a strong contender.

VT-4

Date 2016
Number Produced 17
Weight 52 tonnes (51.2 tons)
Length (with gun) 10.1m (33ft 2in)
Width 3.5m (11ft 6in)
Height 2.4m (7ft 9in)
Engine Norinco 12150Z V12-cylinder turbo-charged diesel, 969kW (1300hp)
Maximum Road Speed 68km/h (42mph)
Crew 3
Armour Modular composite + modular spaced + ERA modules + ECM + side skirts
Main Armament 125mm (4.9in) Type 99 L/80 smoothbore cannon
Main Gun Ammunition Stowed 38

VT-5 (ZTQ-15)

The recently unveiled VT-5 continues the Chinese tradition of developing light tanks. It incorporates many systems proven on recent MBT designs, and will probably replace the Type 62 light tank in service.

follow the general trend of creating a cheaper version of a high-end tank for export. While still supplying clients with a very capable weapon system, export variants generally do not have the same capabilities as the parent version, which helps keep sensitive technologies out of the hands of potential rivals and enemies. Even if another power has its own version of the same system, possession of an example allows its precise capabilities to be determined and countermeasures to be developed.

A little doubt is a powerful, if subtle, weapon, so it is generally preferable that potential enemies cannot be certain they know what they are up against. The extra income from a slightly more capable export tank might help offset the development cost, but the strategic risks are not worth it. Besides, a potential client might not be able to afford a top-end tank and may look elsewhere for a vehicle that fulfils its needs and does not impose such a high maintenance burden.

Supplying export versions to overseas clients has other advantages. Their use in local conflicts will showcase the capabilities of the supplying nation, which can be a useful political tool, while allowing Chinese tank designers to evaluate the performance of their vehicles and plan the next model or even the next generation. It is likely – and certainly it is to be hoped – that the current generation of MBTs will never see action against their equivalents in a major conflict, but without practical experience future developments can be based only on supposition and test data.

In that respect, the tank design world has not changed enormously. Early German panzers were tried out in the Spanish Civil War of the 1930s; Western and Soviet tanks clashed in the Arab–Israeli wars of the 1960s to the 1980s. The next generation of Chinese tanks may well be based upon experience gained by export clients operating MBT-2000s and VT-4s. At present, only Thailand operates the VT-4, but Chinese designs have demonstrated that they can compete with those of any other nation – in financial terms as well as combat effectiveness – and China has emerged as a major player in the armoured vehicles marketplace.

ZTQ-15

Date 2018
Number Produced Not Known
Weight 36 tonnes (35.43 tons)
Length (with gun) 9.2m (30ft 2in)
Width 3.3m (10ft 9in)
Height 2.5m (8ft 2in)
Engine 745kW (1000hp) diesel
Maximum Road Speed 70km/h (43mph)
Crew 3
Armour Modular composite + modular spaced + ERA modules + ECM + side skirts
Main Armament 105mm (4.1in) rifled cannon
Main Gun Ammunition Stowed 38

Type 99A2

Type 99A2
Date 2014
Number Produced Not Known
Weight 58 tonnes (57 tons)
Length (with gun) 11m (36ft)
Width 3.7m (12ft 2in)
Height 2.35m (7ft 8in)
Engine Norinco 12150ZL V12-cylinder turbo-charged diesel, 1119kW (1500hp)
Maximum Road Speed 80km/h (50mph)
Crew 3
Armour Modular composite + modular spaced + ERA modules + side skirts
Main Armament 125mm (4.9in) ZPT98 smoothbore cannon
Main Gun Ammunition Stowed 42

Main Armament
The Type 99 mounts a fully-stabilized ZPT98 125mm (4.9in) smoothbore gun. While a rifled weapon might be more accurate when firing some kinds of ammunition, a smoothbore is more effective when using fin-stabilized munitions and can launch guided missiles. Smoothbores also suffer less wear per shot, giving a longer period between overhauls.

Redesigned turret
The Type 99A2 has a redesigned turret with additional upgrades to its electronic systems and may have a new engine. Details are, as always, difficult to obtain from Chinese sources. The new version is likely to receive the active defence system currently installed on the VT-4. This system uses radar to detect incoming missiles and launches interceptor munitions to shoot them down. Its precise capabilities are largely unknown at this time, but similar interception devices are claimed to be able to intercept mortar bombs in addition to missiles. It is thus possible that Chinese MBTs may be able to intercept indirect-fire munitions as well as missiles approaching more or less parallel to the ground.

Armour
The composition and characteristics of armour are rarely revealed by any tank manufacturer. The Type 99 is known to have advanced composite armour, presumably composed of layered metal and ceramics. Blocks of Explosive Composite Armour on the hull and turret provide additional protection from shaped-charge anti-armour weapons.

Turret
The Type 99A2 received a redesigned turret with a lower overall height and additional armour panels. The prominent bustle at the rear is a feature of most modern tanks, allowing ammunition to be accessible no matter which direction the turret is facing.

Engine
Powerplant technology has made possible China's advanced MBTs. The 36.5-tonnes (36-ton) Type 59 had a 390kw (520hp) engine and a maximum road speed of 50kph (31mph). The Type 99's engine is rated at 1119kW (1500hp) and is capable of driving the 58.9-tonne (58-ton) tank at a top speed of 80m/h (50mph).

2-005

ARMOURED PERSONNEL CARRIERS & INFANTRY FIGHTING VEHICLES

The APC was the final component that allowed the ideal of highly mobile armoured warfare to be realized, and ushered in an era where the 'air-land battle' featuring massed tank assaults backed by mechanized infantry, artillery and air support was the decisive mode of combat. If the emerging armed forces of China wanted to be able to engage in such operations then large numbers of APCs would be needed. In this chapter the following vehicles are featured:

- Type 77
- Type 63
- Type 86
- Type 85/Type 89
- Type 90
- Type 92
- ZBD-97
- ZBD-09 Snow Leopard
- ZBD-2000/ZBD-05

Infantry fighting vehicles (IFVs) like the ZBD-04 can provide fire support to their troops, removing obstacles such as machine-gun posts without needing to call for tank or air support.

Type 77

The Type 77 APC was, like early tank designs, heavily influenced by a Russian vehicle. This was the BTR-50, which was rather basic even when it entered service in 1954.

The Type 77 was not a direct copy, although it was very similar in appearance and general layout. With a driving compartment at the front and the engine at the rear, the troop compartment lay between and was accessed via roof and side hatches.

This arrangement could be inconvenient at the best of times. As experience in World War II showed, troops exiting a vehicle under fire were likely to suffer casualties as they struggled out of the top – especially if enemy personnel knew to expect this and swept the vehicle's roof with fire. Although firing ports were provided for the occupants, firing from a moving vehicle of this vintage has been compared to trying to shoot accurately from inside a blender.

Amphibious APC

The Type 77 was amphibious as standard, a characteristic derived from the Soviet doctrine of maintaining the momentum of an advance despite water obstacles such as rivers. The Type 77 used waterjet propulsion and reportedly displayed good buoyancy characteristics. Armour protection, by virtue of the vehicle's role, had to be light and armament was limited to an unprotected machine gun for use by the crew.

Although its general form was based on the BTR-50, the Type 77 was developed alongside the Type 63 amphibious light tank, which in turn was heavily influenced by the Russian PT-76. The resulting vehicle was very

Type 77
Date 1978
Number Produced Not Known
Weight 15.5 tonnes (15.25 tons)
Length (with gun) 7.15m (23ft 4in)
Width 3.2m (10ft 4in)
Height 2.16m (7ft)
Engine 1250L-2 diesel, 300kW (402hp)
Maximum Road Speed 60km/h (37mph)
Crew 2
Armour 2–13mm (0.07–0.51in)
Main Armament 12.7mm (0.5in) heavy machine gun
Main Gun Ammunition Stowed 500 rounds

Type 77-2
The Type 77-2 was an unsophisticated workhorse vehicle, providing the PLA with lightly armoured transportation and serving in a variety of specialist roles.

Type 77-2

In marine service the Type 77-2 could be expected to make much greater use of its amphibious capabilities.

similar to, but a little more advanced than, the BTR-50, notably featuring a more powerful engine that improved performance. It was adopted for service in 1978, essentially providing the Chinese military with an armoured transport vehicle from the late World War II era.

In addition to its primary role as an APC carrying up to 16 troops, the Type 77 served as an armoured utility vehicle for forward areas, carrying supplies and ammunition and, like many APC designs, the chassis was used as the basis for other vehicles. The first of these, designated Type 77-1, was an artillery tractor designed to carry a disassembled 85mm (3.34in) gun on its roof, with winches to lower it into a firing position.

Type 77-2

The Type 77-2 version, appearing in the early 1980s, was produced in APC, ambulance, artillery tractor and command versions; an armoured recovery vehicle version was also produced. The chassis was used as the basis for the Type 89 self-propelled gun, using a 122mm (4.8in) howitzer derived from the Soviet D-30, and for a non-amphibious platform carrying a surface-to-air missile.

Production of the Type 77 went on until the late 1980s. Capable of providing protection from small-arms fire, the vehicle was entirely suitable for security and transportation tasks that did not require its amphibious capabilities.

However, its main role was to permit marine forces to make an assault landing or advance units of an army force to cross water obstacles and either press on or provide security for heavier vehicles fording, wading or using bridging equipment.

Although now very dated, some examples – estimated at around 200 or so – remain in service with the Chinese armed forces. It is not uncommon for specialist versions of an armoured vehicle such as ambulances or support weapon carriers to remain in service for many years after the main model has been withdrawn. It is not clear if that is the case with the Type 77, but it seems likely that at least some specialist versions will remain in the field for the foreseeable future.

Type 63 (YW531)

The Type 63 APC was in many ways a more conventional vehicle than the Type 77. It was to be amphibious with preparation rather than being able to move in and out of water without delay, and made fewer concessions to buoyancy than the Type 77.

The design process produced a fairly standard APC that proved suitable for conversion to other roles. Nominal crew was two – driver and commander – but up to two additional crewmembers could be accommodated in the driving compartment. Troop capacity is officially listed as 15, but a figure of around 10 seems more likely in practice.

The commander's position had a fully rotating periscope for all-round vision when 'buttoned up' but, unusually, the vehicle's armament was not positioned for the crew to use. Instead, the Type 63 carried a heavy machine gun on a mount accessed by a hatch in the troop compartment. If running without infantry on board, a crewmember could move back to man the weapon, but this might interfere in communication between the commander and driver.

The Type 63 was not as obviously an amphibious vehicle as the Type 77, but could propel itself in water with its tracks. No waterjets or other dedicated water propulsion equipment systems were fitted.

The tracks themselves ran on four large roadwheels, with the drive sprockets located at the front and an idler at the rear. Initially, a 194kW (260hp) diesel engine provided motive power, although a 239kW (320hp) model became available later.

Roof hatches were fitted and could be used for troop movement in and out of the vehicle or to permit individual weapons fire, but for the most part access was by way of the rear doors. This system provided far better protection for troops exiting a vehicle under fire; the armoured front of the APC facing the threat was sufficient to stop heavy machine gun rounds,

Type 63
Date 1963
Number Produced 8000
Weight 12.6 tonnes (12.4 tons)
Length 5.5m (18ft 1in)
Width 3.2m (10ft 5in)
Height 2.52m (8ft 3in)
Engine KHD BF8L 8-cylinder Turbo Diesel 239kW (320hp)
Maximum Road Speed 65km/h (40mph)
Crew 2
Armour Max 14mm (0.55in)
Main Armament 12.7mm (0.5in) Type 54 heavy machine gun
Main Gun Ammunition Stowed 500 rounds

Type 63
This early model Type 63 epitomises the role of the APC – to transport an armoured box full of soldiers to the battle zone.

WZ701 Command Vehicle

The command version of the Type 63 was designated WZ-701. It carried multiple radios and a generator to power them instead of troops.

WZ750 Armoured Ambulance

The WZ-750 field ambulance variant was given a raised ceiling inside the troop compartment, allowing medical personnel to stand as they worked.

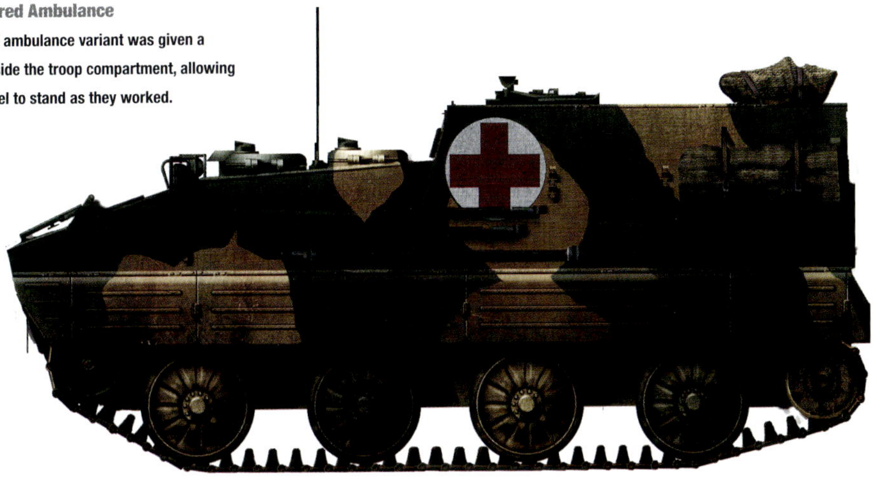

with the sides and other faces offering small-arms protection.

Primary PLA APC

The Type 63 was manufactured in large numbers and was the main Chinese APC until the late 1990s, when it was supplanted by the Type 89. Many variants were produced including the Type 63-I, a general upgrade with improved suspension. Ambulance and command versions were also created, as well as several specialist vehicles based on the chassis. These included a more buoyant version for use by Chinese marines, a self-propelled gun carrier, communications vehicles and a rocket artillery platform.

In addition to the large numbers used by Chinese forces, the Type 63 was also widely exported in several variants. North Korean Type 63s saw action in the latter stages of the Vietnam War and later faced Chinese examples of the same vehicle in the Sino–Vietnamese war. In Iraqi hands, the Type 63 participated in the Iran–Iraq conflict of 1980–88 but proved very vulnerable during the 1991 Gulf War. Later tests indicated that the armour could sometimes be penetrated by small arms, and was prone to spalling even if it was not. As a result, the

WZ303 Multiple Rocket Launcher

Date 1970
Number Produced 400
Weight 12.6 tonnes (12.4 tons)
Length 5.5m (18ft 1in)
Width 3m (9ft 9in)
Height 2.58m (8ft 5in)
Engine KHD BF8L 8-cylinder turbo diesel 240kW (320hp)
Maximum Road Speed 65km/h (40mph)
Crew 3
Armour Max 14mm (0.55in)
Main Armament 19-tube 130mm (5.11in) rockets
Main Gun Ammunition Stowed Not Known

surviving Iraqi vehicles were withdrawn from service, though the Type 63 and its derivatives remain active with other armed forces.

One major change that occurred during the Type 63's long career was the amalgamation of defence industry manufacturers that created the China North Industries Group Corporation, better known as NORINCO. As with its military hardware, Chinese industry was disorganized and mismatched in the early Cold War era, but gradually coalesced into a streamlined whole capable of undertaking large projects in a cost-efficient manner. This was essential to creating a coherent armed force, given the large number of vehicles required, and also permitted entry to the international arms market as a major player. NORINCO is today one of the world's largest defence manufacturers, undertaking development as well as production.

WZ303 MRL

Using the Type 63 chassis for the WZ-303 artillery variant avoided the necessity of constructing a new vehicle from scratch.

Type 63-2

Half-turret protection for the machine gun operator was introduced on later model Type 63 APCs. It is rarely seen elsewhere.

Type 86 (WZ501)

The armoured personnel carrier was developed to meet the needs of highly mobile warfare, allowing infantry to keep pace with the armoured forces and to operate in a mutually supporting role.

In order to achieve this, the infantry needed a 'battle taxi' capable of providing protection from small-arms fire at the least. Most APCs are armed with a machine gun for self-defence, and although this can be used to support dismounted infantry, a more potent capability was desirable.

The result was the infantry fighting vehicle (IFV). Carrying weapons capable of eliminating many threats to its infantry contingent, an IFV adds to the fighting power of a force rather than simply providing transportation and may carry anti-tank or anti-aircraft weapons to provide additional capabilities. Indeed, IFVs are sometimes considered an alternative to battle tanks, especially by militaries that do not expect to face them.

To be effective, an IFV has to operate in proximity to the enemy, who may well have anti-vehicle weapons. This creates a requirement for greater protection, which adds weight; this in turn increases power and fuel requirements. The end result of all this is that an IFV is vastly more expensive than an APC with the same troop capacity.

Nevertheless, the militaries of the major powers began to create such vehicles once the concept was demonstrated. The title of first true IFV is normally claimed by the Soviet BMP-1, although other vehicles have been put forward. The BMP-1 is generally given the credit since it mounted a 73mm (2.9in) gun and an anti-tank missile launcher, giving it significant fighting power against armoured and unarmoured opponents.

New Chinese IFV

By the time China decided to develop an IFV of its own, the BMP-1 had already been in service for a decade and a half and had enjoyed enormous export success. This made it relatively easy to obtain a few examples through third parties, which was necessary since relations with Russia were not good enough at the time to make direct purchase a possibility. The

Type 86

Chinese designers took a low-risk approach by copying the Russian BMP-1 to create their first IFV, the Type 86.

Type 86
Date 1986
Number Produced At least 1000
Weight 13.3 tonnes (13.8 tons)
Length 6.74m (22ft 1in)
Width 2.97m (9ft 9in)
Height 2.16m (7ft)
Engine Type 6V-150 diesel, 222kW (298hp)
Maximum Road Speed 65km/h (40mph)
Crew 3
Armour 2–13mm (0.07–0.51in)
Main Armament 1 x 73mm (2.9in) smoothbore cannon; Red Arrow 73 ATGW launcher mounted on main gun
Main Gun Ammunition Stowed 40 main gun; 5 missiles

BMP-1 was then reverse-engineered and a Chinese version developed under the designation WZ-501.

Entering service in 1992, the Type 86 retained most of the features of the BMP-1, not all of them positive. The troop compartment, capable of holding eight personnel, was cramped and normally accessed through the rear doors – which contained fuel tanks. The four roof hatches provided alternative access, though with the usual problems of exiting through the top of a vehicle in combat, and could be locked in the open position to provide protection to troops standing up to fire their personal weapons.

Arms and armour

Protection all round was sufficient to defeat small arms, with the front of the vehicle proof against heavy machine guns. The primary armament, a smoothbore 73mm (2.9in) cannon similar to the original Russian weapon, was inaccurate and suitable mainly for providing fire-support against 'soft' targets using HE-frag ammunition. HEAT rounds were available, but the unstabilized gun was unlikely to hit a moving target even from a stationary position, and firing on the move was pointless.

The 73mm (2.9in) gun was carried in a low turret that made a small target. However, this was offset by the large overall size of the vehicle and the very limited depression of the gun. It is not possible to gain a concealed or protected hull-down position, so if a Type 86 was to support its infantry contingent it must expose itself to

Type 86 IFVs were among the Chinese armoured vehicles taking part in the Vostok 2018 exercises alongside their Russian hosts.

fire – and it could be penetrated at the sides and rear by anything heavier than 7.62mm (0.3in).

The HJ-73 'Red Arrow' missile system offered some anti-armour capability, but the Type 86 was no tank hunter in its basic configuration. The HJ-73 was China's first-generation anti-tank missile, largely copied from the Russian 9M14, better known to Western observers as the AT-3 Sagger. It used semi-active command line-of-sight (SACLOS) guidance, requiring the gunner to guide the missile all the way to the target. This is difficult to do from a moving vehicle, so the Type 86 could really only strike against armoured vehicles from ambush. The slow-flying HJ-73 could reach a range of 3000m (3280yd), but took around 25 seconds to do so.

Thus, China's first IFV was based on an outdated design and offered relatively slight advantages over a typical APC. By comparison, the US M2 Bradley, a greatly more capable vehicle designed at least in part to counter Soviet BMPs, entered service a decade earlier. However, the Type 86 allowed the Chinese armed forces to gain experience of IFV operations, and was the basis for a wide range of variants.

Variants

In addition to command and ambulance versions, and a variant optimized for marine operations, the

Type 86

A Type 86 in Chinese marine camouflage. Deploying the trim vane at the very front of the hull and starting the bilge pumps would prepare the marine type 86 for amphibious operations.

Type 86B

Modifications improving the vehicle's performance in water created the Type 86B, which is in service with the PLA marine forces.

A Chinese Type 86 IFV competing in the 2016 International Army Games, in a challenge testing both offroad and offensive capabilities.

Type 86A (WZ501A)
Date 1986
Number Produced Not Known
Weight 13.3 tonnes (13.8 tons)
Length (with gun) 6.74m (22ft 1in)
Width 2.97m (9ft 9in)
Height 2.16m (7ft)
Engine Type 6V-150 Diesel, 222kW (298hp)
Maximum Road Speed 65km/h (40mph)
Crew 3
Armour 2–13mm (0.07–0.51in)
Main Armament 1 x 20mm (0.8in) cannon, 7.62mm (.03in) coaxial MG

Type 86 was the basis for various experimental tank-hunters. The concept of such vehicles is sound, and similar to the thinking behind gun-armed tank destroyers. A force of IFVs armed with missiles should – under the right conditions – be able to eliminate enemy tanks for a fraction of their cost.

Of course, such vehicles were extremely vulnerable if the odds were not heavily stacked by an ambush, and lacked the versatility of tanks. Experimental vehicles with multiple missile launchers were trialled, and appear to have been at least marginally effective.

Modern upgrades

Modernized versions of the Type 86 were equipped with more effective weaponry. The smoothbore 73mm (2.9in) gun was replaced with a light autocannon that offered improved support-fire capabilities against a range of targets. Locally produced guns were generally used, but a short-lived US/Chinese joint project

Type 86A

The Type 86A variant replaces the 73mm (2.9in) smoothbore gun with a light autocannon in a remotely operated turret, or a heavy machine gun on export models.

Type 86G
Date 1986
Number Produced Not Known
Weight 13.3 tonnes (13.8 tons)
Length (with gun) 6.74m (22ft 1in)
Width 2.97m (9ft 9in)
Height 2.16m (7ft)
Engine Type 6V-150 diesel, 222kW (298hp)
Maximum Road Speed 65km/h (40mph)
Crew 3
Armour 2–13mm (0.07–0.51in)
Main Armament 1 x 30mm (1.2in) autocannon; Red Banner ATGM HJ-73
Main Gun Ammunition Stowed 500 main gun; 5 missiles

produced a version armed with a US 25mm (1in) Bushmaster cannon.

Around 1000 Type 86 IFVs were built, serving with both the Chinese army and marines. Export sales are hard to verify, not least since the very similar BMP-1 is in service with several nations. Iran based its own indigenous IFV on the Type 86. This vehicle, designated Boragh, is still in service and has been developed into variants including mortar, missile and howitzer armed vehicles, as well as a light air defence platform.

Type 86G
The Type 86G uses a 30mm (1.18in) cannon in a one-man turret developed for the ZLC-2000 airborne combat vehicle.

A Chinese ZBD-86A infantry fighting vehicle takes part in trials during the final stage of the Suvorov Attack competition at the Alabino Firing Range as part of the 2016 Army Games, an international event organized by the Russian Defense Ministry.

Type 85/Type 89

The project that eventually produced the Type 85 and Type 89 was intended to create a more capable replacement in the APC role, which could be expanded into an IFV if needed.

Its predecessor, the Type 63 was a fairly modest vehicle, capable of functioning as a 'battle taxi' but lacking any real capacity to support its troop contingent. This of course must be balanced against cost and overall simplicity. The initial model was designated Type 85, and was offered for export. A developed version, with a slightly larger hull, was taken into Chinese service with the designation Type 89.

The basic concept for the new vehicle was an enlarged Type 63 featuring NBC protection for the occupants. The vehicle was to be amphibious with preparation. To support the lengthened hull, an extra roadwheel was added, while a larger and more powerful engine, combined with improved transmission, improved mobility and reliability. In addition to the crew of two, up to 15 infantrymen could be carried. Alternatively, the rear compartment could be used to carry a variety of weapon systems in the support and IFV roles.

Armour

The designers of the Type 85/89 chose a fairly standard level of protection. The vehicle was designed to defeat small-arms fire and shell fragments from all directions, and to be able to withstand heavy machine gun fire from the front. This was, on paper, no better than the Type 63 but the armour was better constructed and implemented as a result of experiences with export versions in the Gulf War.

It is never possible to give an APC or IFV the same level of protection as a tank. The requirement to carry multiple infantry soldiers creates a large space that must be roughly box-shaped, making it impossible to slope armour to any significant degree. This volume must be given some protection, but an APC or IFV simply could not carry the weight required to enclose it in slabs of

Type 85 (YW531H)

The Type 85 was created by scaling up the Type 63, adding an extra roadwheel to support the lengthened hull. It was designated YW531H in Chinese service.

Type 85 (YW309)
Date 1985
Number Produced 358
Weight 14.7 tonnes (14.5 tons)
Length 6.27m (20ft 7in)
Width 3.06m (10ft 5in)
Height 2.48m (8ft 2in)
Engine Deutz BF8L413F air-cooled diesel, 239kW (320hp)
Maximum Road Speed 65km/h (40mph)
Crew 3 (+ 8 troops)
Armour RHA 5–14mm (0.2–0.6in)
Main Armament 73mm (2.9in) smoothbore short-recoil semi-automatic cannon; Red Arrow 73 ATGW launcher mounted on main gun
Main Gun Ammunition Stowed 40

composite armour. Thus most APCs are quite lightly armoured, and the Type 89 was no exception.

Armament on the APC version was likewise standard; a heavy machine gun was mounted on the roof. A manually operated weapon of this sort has the advantages of cheapness and simplicity, but it requires a crewmember to expose himself to enemy fire in order to operate it. This problem was addressed by the addition of an unusual shield for the gunner, encasing him to the sides and rear in light armour.

Command and ambulance versions of the Type 89 were relatively easy to produce, requiring little alteration to the vehicle as a whole. The command vehicle uses much of its internal space

Type 85 (YW309)
Adding the turret of a Type 86, containing a 73mm (2.87in) smoothbore gun and missile launch system, converted the Type 85 into an IFV.

Type 85 (WZ504)
The tank hunter variant was limited by the capabilities of its four Red Arrow 73 anti-tank guided missiles, but could be effective under the right circumstances.

for communications equipment and has no machine gun, while the ambulance retains its weapon. It also has a higher ceiling in the rear compartment to allow medical personnel to work.

Specialist variants

Various infantry-fighting vehicle and specialist variants of the Type 89 also emerged. One IFV was created by adding the turret of the Type 86 to a Type 89. This vehicle, internally designated YW309, had the advantage of simplicity but the vehicle it created was no more capable than the Type 86. The limitations of the smoothbore 73mm (2.9in) gun and the Red Arrow missile system remained. A more modern-appearing and probably more capable version was designated YW307. This mounted a 25mm (1in) automatic cannon and a co-axial machine gun. It is difficult to compare the effectiveness of the gun and cannon, but overall an automatic weapon was more likely to put some rounds into or close to the target than the highly inaccurate 73mm (2.9in) smoothbore.

The addition of a turret reduced troop complement to seven rather than the 15 of the APC version. This is one of the trade-offs of using IFVs – more vehicles are needed, and those vehicles are more expensive than standard APCs. Yet they are also more capable and can contribute a great deal to the overall effectiveness of the force. The question is really one of affordability. One way to achieve this is to get the most out of a given chassis by using it for a whole family of vehicles. That way, development costs are essentially shared across a range of roles.

In the case of the Type 89, a number of artillery vehicles were created. Mortar carriers, armed with 120mm (4.7in) and 82mm (3.2in) weapons,

The ZSD-89-II variant is in service with the armed forces of Zimbabwe and, as pictured here, Sri Lanka.

Type 89

The 12.7mm (0.5in) machine gun is used, with or without a shield, on most variants of the Type 89 that are not IFVs equipped with a turret.

Type 89

The Type 89 has more and smaller roadwheels than its predecessor, with the tracks driven by a front sprocket.

Type 89
Date 1985
Number Produced 358
Weight 14.7 tonnes (14.5 tons)
Length (with gun) 6.27m (20ft 7in)
Width 3.06m (10ft 5in)
Height 2.48m (8ft 2in)
Engine Deutz BF8L 413F 4-stroke air-cooled diesel, 239kW (320hp)
Maximum Road Speed 65km/h (40mph)
Crew 3 (+ 8 troops)
Armour RHA 5–14mm (0.2–0.6in)
Main Armament Type 14 12.7mm (0.5in) heavy machine gun
Main Gun Ammunition Stowed 1120

were likely the easiest to implement. The high trajectory of a mortar and its small size permit one to be mounted inside a vehicle with few modifications. Artillery weapon platforms such as the YW323 with a 122mm (4.8in) gun and the PLZ-45, equipped with a 155mm (6.1in) gun-howitzer, require much greater modification to the vehicle. A multiple rocket launcher variant was also produced, along with a missile-armed tank-hunter. In addition to these combat vehicles, the Type 89 was also the basis for armoured recovery and engineering vehicles as well as logistics platforms and a mine-laying vehicle.

Some variants use a remotely controlled weapon in an unmanned turret, a feature that has been seen on armoured vehicles of various nationalities in recent years. An unmanned mounting can be much smaller than a turret, and might be left all but unarmoured as the weapon is unlikely to be hit. Even where armour is provided, the volume that needs to be protected is much smaller. In addition to saving overall weight, the lack of a bulky armoured turret can also improve stability when tackling adverse slopes.

YW307

Date 1989
Number Produced 358
Weight 15.4 tonnes (15.1 tons)
Length (with gun) 6.15m (20ft 2in)
Width 3.13m (10ft 3in)
Height 2.93m (9ft 7in)
Engine Deutz BF8L 413F 4-stroke air-cooled diesel, 239kW (320hp)
Maximum Road Speed 65km/h (40mph)
Crew 2 (+7 troops)
Armour RHA 5–14mm (0.2–0.6in)
Main Armament 25mm (1in) ZPT-90 autocannon
Main Gun Ammunition Stowed 1200

YW307

Light autocannon, in this case 25mm (1in), are perhaps the best all-round weapons for an IFV, offering the ability to engage a variety of targets, including light armoured vehicles.

ZDS-89-II

The current version of the Type 89 in Chinese service is designated ZSD-89-II. As with other Chinese vehicles, sources vary about the exact designation.

WZ534

Some export versions of the Type 89 were given an enlarged hull, with an additional roadwheel to support it.

Type 90 (WZ551)

Wheeled vehicles offer many advantages over tracked versions, at least in some circumstances. They are easier to build and maintain, do less damage to roads and do not require specialist vehicles to move them over long distances.

This gives wheeled vehicles greater strategic mobility under their own power, although a tracked version can often cross terrain that would be impassable to anything relying on wheels.

Wheeled vehicles also send a different message when they are deployed. Tracked APCs and IFVs are often misidentified as 'tanks' in the media; even if they are not, the deployment of tracked combat vehicles has 'warfighting' connotations, whereas a wheeled APC may be seen as a 'security' asset. Whether or not the public-relations dimension mattered to the Chinese military, the development of wheeled armoured transport was seen as sufficiently important that the WZ523 entered trials in 1984.

The WZ523 was a six-wheeled light APC capable of carrying 10 infantrymen plus its crew of two. Although the concept was sound, the vehicle itself was not adopted for service. Further work produced the WZ551, which entered service under the designation Type 90. This vehicle was also unsatisfactory and was produced only in small numbers.

Light wheeled APC

The main problem with the WZ551 was a lack of engine power. To remedy this, an upgraded version was developed jointly by the NORINCO and GIAT. Sometimes designated NGV-1, this version featured evenly spaced wheels rather than the one-and-two configuration of the previous model. The sole armament of the original Type 90 design, a roof-mounted heavy machine gun, was replaced with a 25mm (1in) cannon. This, arguably, put the NGV-1 in the IFV class as its main weapon could now penetrate most lightly armoured vehicles, including Chinese and Soviet APCs.

The collaboration with GIAT fell victim to international politics, cutting off the supply of guns that were to have armed the NGV-1. A virtually identical weapon was developed for local manufacture, following a long Chinese tradition of copying overseas technologies whether or not permission was granted.

The end result of this process was a light APC suitable for security and fast-reaction forces, capable of deploying 10 infantrymen and providing them with fire-support. Equivalent vehicles were appearing worldwide, reflecting changes in both the armoured vehicles marketplace and the combat environment that drives it. In the years immediately after World War II, the focus was on large-scale conflict fought by tanks and mechanized infantry, with

Type 90

Date 1995
Number Produced More than 2200
Weight 12.5–15 tonnes (12.3–14.7 tons)
Length (with gun) 6.63m (21ft 9in)
Width 2.8m (9ft 2in)
Height 2.89m (9ft 6in)
Engine Deutz BF8L413F 8-cylinder turbo diesel 235kW (320hp)
Maximum Road Speed 85km/h (53mph)
Crew 3
Armour Not Known
Main Armament ZPT90 25mm (1in) cannon (IFV version)
Main Gun Ammunition Stowed Not Known

heavy air support. It was becoming apparent by the 1980s that although the capability to fight a superpower conflict had to be maintained, there was a need for equipment tailored to lower-intensity conflicts.

Using main battle tanks against a primarily insurgent force was a lot like taking on a swarm of ants with a sledgehammer. There might be occasions where the concentrated fighting power of tanks would be useful, but on the whole any vehicle that was resistant to rifle fire and mounted a decent anti-personnel weapon would be sufficient.

Type 90 (WZ523)

The WZ523 was an effective light APC well suited to low-intensity conflict, convoy escort and security operations.

Type 90 (WZ551)

The WZ551, or Type 90, proved to be underpowered and was not a success. Planned variants were cancelled and re-implemented on the chassis of its successor.

The new generation of wheeled armoured vehicles ranged from the US high-mobility multipurpose wheeled vehicle (HMMWV), which entered service in 1985, to the Russian BTR-80 of 1987. While taking different approaches to the problem of equipping a force for global security rather than warfighting operations, these designs all faced the same challenges. They needed to provide mobility and enough firepower to give troops a clear advantage over their opponents at a price that allowed sufficient coverage for effective operations. Vehicles like the Type 90 and its overseas equivalents were sufficient for this role, and have on the whole seen more active service than the heavy armoured spearhead forces. However, the shortcomings of the Type 90 meant that it was quickly supplanted by an improved version designated Type 92. The planned specialist vehicles that followed the Type 90 into service were thus based on its replacement instead.

Type 92 (WZ551A)

Developed from the NGV-1, the Type 92 was visually very similar but of heavier construction. Prototypes began evaluation in 1991, with the basic vehicle entering service in 1995.

Designated WZ551A or Type 92A, the APC model was equipped with a 12.7mm (0.5in) machine gun, while the Type 92B model might be considered a light IFV and was armed with a 25mm (1in) cannon. Both versions carried nine infantry personnel in addition to their crew of three.

Using a 6x6 wheeled configuration, the Type 92 was designed to be fully amphibious. Propulsion in water was by means of two small propellers at the rear of the vehicle. On level ground, the Type 92 can reach a top speed of around 85km/h (53mph), significantly more than a tracked vehicle. An operating radius of 800km (497 miles) also offered greater strategic mobility than a tracked APC with similar capabilities.

Type 92A

The basic Type 92 APC has a nominal capacity for nine infantrymen, who use four roof hatches and a rear door to access their compartment.

Type 92A (WZ551A)

Date 1995
Number Produced 2015 (+ 209 exported)
Weight 12.9 tonnes (12.3 tons)
Length (with gun) 6.63m (21ft 9in)
Width 2.8m (9ft 2in)
Height 2.8m (9ft 2in)
Engine Deutz BF8L 413F diesel, 239kW (320hp)
Maximum Road Speed 85km/h (53mph)
Crew 3 (+ 9 passengers)
Armour RHA
Main Armament 12.7mm (0.5in) QJC88 heavy anti-aircraft machine gun
Main Gun Ammunition Stowed 1500

WZ551A Command Vehicle

The command variant of the Type 92 is recognizable by its enlarged rear compartment. It appears that command APCs are not routinely armed.

WZ550 4x4 Tank Hunter

The tank hunter variant carries four short-range HJ-9 ATGM missiles. However, its mobility may enable it to strike from ambush then escape before a response can materialize.

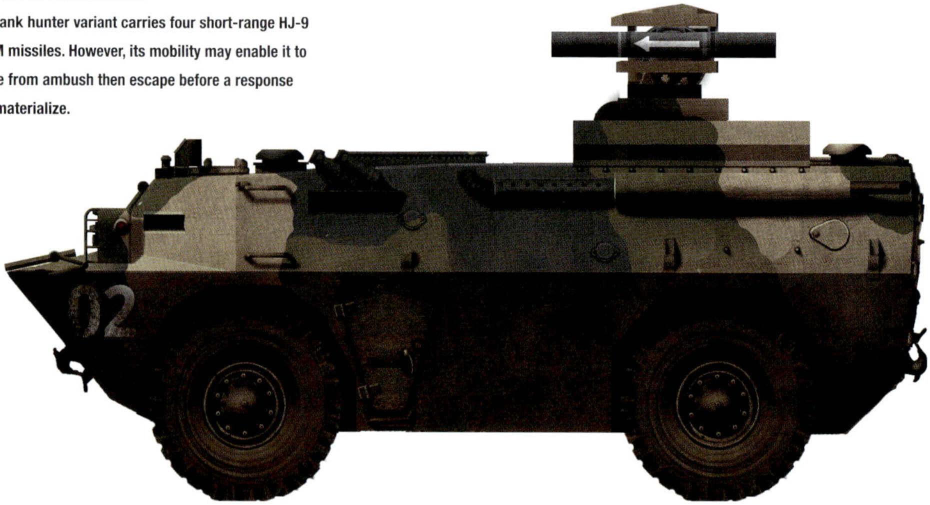

Multiple capabilities

The Type 92 was armoured against small arms and heavy machine guns, and provided NBC protection to the occupants. In addition to its APC role, the Type 92 was capable of carrying a range of armament.

The IFV version, 1.5 tonnes (1.45 tons) heavier than the APC, carried a 25mm (1in) cannon. On some variants of this vehicle the cannon used a high-elevation mount, perhaps to provide air defence against helicopters. A 30mm (1.2in) cannon may have been substituted on some examples.

Minor variations and local modifications can make it difficult to distinguish between the Type 90 and Type 92, such as this example of the Type 92A in Sri Lankan service.

Like many wheeled armoured vehicles, the standard 6x6 APC was supplemented by 4x4 and 8x8 versions suitable for use as armoured cars, reconnaissance vehicles and weapons carriers. Although requiring a different hull and chassis, these vehicles had many parts in common with the standard APC and IFV, simplifying logistics and crew training.

Among the derivative vehicles were several self-propelled artillery platforms using mortars or tube artillery. Support platforms based on the same vehicle as the infantry use can keep pace and cross the same terrain, and of course can share maintenance facilities.

A 'tank destroyer' version armed with a smoothbore 100mm (3.9in) gun was also implemented under the designation PTL02 or Type 02. Despite its name, the PTL02 was not effective against modern main battle tanks, but its gun could penetrate most tanks of an older generation as well as lighter vehicles. It also had fire-support applications against bunkers or buildings. Indeed, this vehicle might be better thought of as a direct-fire-support platform with the capability to engage lightly armoured vehicles.

Export success

As a relatively inexpensive APC or weapons carrier, the Type 92 achieved some success on the international market. Many potential clients lack the skills required to maintain tracked vehicles or have requirements that make them unsuitable. Users include Bosnia, Kenya, Sri Lanka and Iran, with additional clients operating the WMA301. This is an export version of the PTL02, armed with a 105mm (4.1in) gun compatible with NATO ammunition. Such compatibility is necessary for export success; in the polarized Cold War world, clients would buy their vehicles and ammunition from one bloc or the other, but today the market is much more open. Clients will shop around for the best vehicle to suit their needs, and being unable to use the industry-standard ammunition might well be a deal-breaker.

The Type 05 mortar carrier is an example of an unsuccessful export version that was subsequently adopted for service with its parent country's armed forces. Armed with a 120mm (4.7in) rifled gun-mortar – which is not compatible with conventional 120mm (4.7in) mortar ammunition – the vehicle was designed to provide highly accurate short-range fire-support but provoked little interest when offered for international sale. This was apparently due to the availability of a very similar Russian vehicle at a lower price.

Type 92B

Date 1995
Number Produced 376
Weight 13 tonnes (12.4 tons)
Length (with gun) 6.63m (21ft 9in)
Width 2.8m (9ft 2in)
Height 2.8m (9ft 2in)
Engine Deutz BF8L 413F air-cooled diesel, 235kW (320hp)
Maximum Road Speed 80km/h (50mph)
Crew 3 (+ 9 troops)
Armour RHA
Main Armament 25mm (1in) ZPT-90 autocannon
Main Gun Ammunition Stowed 400

Type 92B

Successful systems, such as the remotely controlled turret mounting a 25mm (1in) cannon, are often used on multiple vehicles, complicating the task of identifying new types or variants.

ZLS-92B

The Type 92B has continued to be modernized and upgraded. This version, armed with a 20mm (0.78in) cannon and anti-tank missiles, is designated ZLS-92B.

ZBD-97

The ZBD-97, or Type 97, entered service in 2006. It was developed to fulfil the IFV and fire-support vehicle role, carrying seven troops in addition to its crew of three. Firing ports are provided for use by the infantry contingent, although the usefulness of this feature remains debatable.

The Type 97 is fully amphibious, and is capable of swimming ashore from an assault ship. It is propelled by two waterjets with a reported speed in water of 20km/h (12mph). A wave-fence prevents the front of the vehicle – and the driver's vision devices – from being swamped during a water transit. On land, a top speed of 65km/h (40mph) is reported, with an operating radius of about 500km (310 miles).

Heavy fire-support vehicle

The ZBD-97 has obvious similarities to the Russian BMP-3, especially in the design of its turret. This mounts a stabilized 100mm (3.9in) gun capable of tackling most targets but that is ineffective against an MBT when firing its standard HE-frag ammunition. For heavily armoured targets or helicopters, the gun can launch laser-guided missiles.

Like many Chinese weapon systems, the missiles launched by the ZBD-97 are based on a Russian weapon – in this case the 9M117 'Bastion' missile. Capable of engaging a moving target out to 4000m (4374yd), this missile follows a laser designator that must be kept focused on the target. Doing so from a moving vehicle is extremely difficult, if not impossible, so the ZBD's capabilities as a tank-destroyer depend on being able to strike from ambush.

ZBD-97
Date 2005
Number Produced 438
Weight 20.2 tonnes (19.9 tons)
Length (with gun) 7.15m (23ft 6in)
Width 3.2m (10ft 6in)
Height 2.5m (8ft 2in)
Engine Deutz BF8M1015CP diesel, 440kW (590hp)
Maximum Road Speed 65km/h (40mph)
Crew 3 (+ 7 troops)
Armour Aluminium/steel RHA
Main Armament 100mm (3.9in) rifled autocannon
Main Gun Ammunition Stowed 30

ZBD-97
The ZBD-97, or Type 97, is deployed with the armoured forces of Guangzhou and Nanjing military regions, where it serves as an amphibious fire-support platform.

In addition to the 100mm (3.9in) gun, the ABD-97 mounts a 30mm (1.2in) cannon, giving it very significant infantry support capability. Automatic cannon of this sort are typically provided with a mix of armour-piercing and HE-frag ammunition to permit the widest range of targets to be effectively engaged. Both the main gun and the autocannon are supported by laser rangefinding, stabilization and advanced fire control electronics.

Like many vehicles in the same class, it is armoured against small arms and shell fragments, and has a collective NBC protection system for the crew and passengers.

Although the ZBD-97 does not mount an active protection system like that of the visually similar BMP-3, it does have laser warning systems and a countermeasures package.

Limited service

The ZBD-97 is in current service with some Chinese armoured formations, and was the basis for an armoured recovery variant. It does not appear to have been built in very large numbers, or have been offered for export.

The current designation of the Type 97 appears to be ZBD-04. Sources are unclear about differences between this model and the ZBD-97, if any.

ZDB-9 Snow Leopard

During the 1990s, various nations began developing large wheeled armoured vehicles that could serve in the APC or IFV role, and that could be developed into weapon carriers, tank destroyers and air defence platforms. A number of 8x8 wheeled armoured vehicles emerged, including the ZBD-09 Snow Leopard, which was officially presented to the world in 2009.

Like many other 8x8 vehicles, the Snow Leopard is a scaled-up version of a 6x6 vehicle, allowing it to carry greater loads and heavier weapons. The additional wheels also increase survivability in the event of encountering a mine or an improvised explosive device. The ability to limp away from the attack site might be vital if mines have been laid as part of an ambush, and in any case being immobilized on a remote patrol route places the vehicle and crew in serious danger.

The original 6x6 version had a number of interesting features, including a modular construction with each major system – engine, controls, suspension and so forth – integrated as a module and then combined to create a complete vehicle. If was offered for export but did not achieve any sales. This did not deter the designers from pressing on with the 8x8 version, which entered service with the Chinese army.

Rapid-reaction IFV

The Snow Leopard family of vehicles is used by rapid-reaction forces, providing armoured transportation in the IFV role. Fire-support, anti-tank and other specialist tasks are carried out by variants of the same vehicle, greatly simplifying maintenance and logistics support. It is probable that all vehicles in the family are fitted with a battlefield management system, allowing information and orders to be shared rapidly throughout the entire force and support called in where it is most needed.

The standard version of the Snow Leopard is the ZBD-09 IFV, equipped with a 30mm (1.2in) cannon in a small turret that could mount rails for

ZBL-08

Eight-wheeled light armoured vehicles are becoming increasingly popular in many nations. They offer a good balance of cost, mobility and the capability to carry effective weaponry.

ZBL-08
Date 2008
Number Produced 200
Weight 15 tonnes (14.6 tons)
Length (with gun) 8m (26ft 3in)
Width 3m (9ft 9in)
Height 2.1m (6ft 10in)
Engine Deutz BF6M 1015C diesel, 328kW (440hp)
Maximum Road Speed 100km/h (62mph)
Crew 3 (+ 7–10 troops)
Armour RHA + add-on modular
Main Armament 12.7mm (0.5in) machine gun
Main Gun Ammunition Stowed 3000

---

launching anti-tank missiles. The IFV can carry 7–10 infantry personnel and has a central tyre pressure regulation system to permit the vehicle to cross very soft ground. It is fully amphibious, using waterjets for propulsion.

Armour protection against small arms in all arcs and heavy machine gun fire from the front is standard, and the APC version of the Snow Leopard probably does not carry any more than this. Ceramic composite armour is available in the form of add-on blocks, which – Chinese sources claim – render the vehicle impervious to 25mm (1in) fire in the frontal arc and 12.7mm (0.5in) elsewhere.

APC version

An APC version, with no turret and armed only with a heavy machine gun, is primarily used for supply and support tasks, with the IFV version confronting the enemy. For direct fire-support, a variant designated ZTL-09 and armed with a rifled 105mm (4.1in) gun is available. Indirect fire can be provided by a mortar armed version (PLL-09) or a self-propelled howitzer (PLZ-09). This vehicle can carry a 122mm (4.8in) or 155mm (6.1in) gun.

An air defence variant, usually armed with a six-barrel 30mm (1.2in) Gatling weapon, completes the range of combat vehicles a mobile force needs: infantry transport, anti-tank, fire-support and air defence. Additional capabilities are provided by the APC version acting as a supply and utility vehicle, and by specialist ambulance, command and armoured recovery vehicles.

The concept embodied by this force is very different from the armoured battlegroups of World War II or the early Cold War. Then, the emphasis was on tanks as a heavy spearhead, smashing holes in an enemy line and exploiting the breakthrough against clearly defined targets. In the modern world, a war (or 'war-like situation') is likely to be far more fluid. A major confrontation against an advanced power fielding large numbers of tanks is possible, but more often military forces are required to chase down insurgents or secure an area against them.

A militia using civilian trucks equipped with machine guns is a very different threat than an armoured battlegroup, and requires a different response. A tank force is invulnerable to attack by such a militia, but it is unlikely to be able to force a decisive battle, and meanwhile the highly mobile enemy would strike targets the tanks could not reach in time. An armoured force could become exhausted chasing around after mobile

ZBD-09 Snow Leopard
Date 2009
Number Produced 537
Weight 19.1 tonnes (18.8 tons)
Length (with gun) 8m (26ft 3in)
Width 3m (9ft 9in)
Height 2.1m (6ft 10in)
Engine Deutz BF6M 1015C diesel, 328kW (440hp)
Maximum Road Speed 100km/h (62mph)
Crew 3 (+ 7–10 troops)
Armour RHA + add-on modular
Main Armament 30mm (1.2in) ZPT-99 (2A72) autocannon
Main Gun Ammunition Stowed 650

ZBD-09
The 30mm (1.2in) cannon of the ZBD-09 can penetrate light armoured vehicles and poses a serious threat to most targets other than tanks.

targets, suffering operational losses and breakdowns that drained its capabilities despite not firing a shot.

The lightly equipped, highly mobile reaction force has an important role to play in the 21st-century military environment. Protection against machine guns and mines is, for the most part, sufficient: having enough vehicles to cover the ground, and being mobile enough to reach a fleeting target, are more important in some situations than raw fighting

ZTL-09

The 105mm (4.1in) rifled gun of the fire-support/assault gun variant is capable of penetrating most of the tanks it might encounter.

ZTL-11

The ZTL-11 amphibious fire-support vehicle is sometimes referred to as ZTL-09. The latter may have been an interim designation used during development from the ZBD-09.

ZTL-09

Date 2009
Number Produced Not known
Weight 20 tonnes (19.6 tons)
Length (with gun) 9m (29ft 6in)
Width 3m (9ft 9in)
Height 3.2m (10ft 5in)
Engine Deutz BF6M 1015C diesel, 328kW (440hp)
Maximum Road Speed 100km/h (62mph)
Crew 4
Armour RHA + add-on modular
Main Armament 105mm (4.1in) rifled gun; coaxial 12.7mm (0.5in) machine gun
Main Gun Ammunition Stowed 30

power. Despite experiments with innovative weapons, such a force cannot realistically expect to defeat an army equipped with modern tanks, but older vehicles are vulnerable to the lighter weapons carried by wheeled tank-killers.

Export success

The IFV and anti-tank versions of the Snow Leopard family have achieved some export success to date. It may be that many nations that were previously in the market for older tanks being sold on by their initial users might instead consider cheaper, easier-to-maintain wheeled vehicles capable of fulfilling the same role. It is unlikely that wheeled vehicles like the ZBD-09 will replace tanks, but they offer an alternative in some situations.

ZBD-09

China is one of the nations to undertake extensive experiments with digital camouflage patterns. Experts remain divided on whether or not these offer any real advantages over traditional camouflage systems.

ZBD-2000/ZBD-05

The ZBD-2000 was developed, like many modern armoured platforms, as part of a family of vehicles. Its primary role is to deliver troops to a beach from amphibious assault ships located up to 10km (6 miles) offshore.

The ZBD-2000, sometimes referred to as ZBD-05, reflects a growing emphasis on expeditionary warfare. The US Marine Corps was to have received a similar vehicle, but this was cancelled for budgetary reasons. China, on the other hand, perceives a need for highly capable amphibious vehicles and is willing to spend the large sums of money required to develop them.

The ZBD-2000 has a reported speed of 25km/h (16 miles) using waterjet propulsion, while on land its road speed is 65km/h (40mph) – comparable with most other Chinese armoured vehicles.

Infantry Fighting Vehicle

The IFV version can carry six to 10 troops, who deploy through the rear doors or roof hatches. They are supported by a turret-mounted 30mm (1.2in) cannon capable of firing armour-piercing or high-explosive ammunition. This weapon appears to have been developed from a Ukrainian design, and like similar weapons is capable of engaging lightly armoured vehicles out to 1500m (1640yd) and 'softer' targets to a much greater range.

Missile platform

Rails for launching HJ-73 missiles are fitted along the sides of the turret, giving a limited anti-tank capability. The long flight time and need for manual guidance makes

ZBD-05

The ZBD-2000/ZBD-05 is variously designated a light tank and an IFV, depending on its armament and role. Both are built on the same hull.

ZBD-05

A ZBD-05 IFV in PLA marine service, wearing maritime digital camouflage. China uses a great variety of camouflage patterns as its forces have to operate in a huge variety of environments, even without deploying overseas.

ZBD-05

Date 2005
Number Produced 231
Weight 26.5 tonnes (26.1 tons)
Length (with gun) 9.5m (31ft 2in)
Width 3.36m (11ft 1in)
Height 3.04m (10ft)
Engine Deutz diesel, 410kW (550hp)
Maximum Road Speed 65km/h (40mph)
Crew 3 (+10 troops)
Armour RHA
Main Armament 30mm (1.2in) ZPT-99 autocannon
Main Gun Ammunition Stowed 600

ZTD-05

The light tank version is not much better armoured than the IFV – some sources state it has exactly the same protection – and is a tank only by virtue of its 105mm (4.1in) armament.

these missiles less effective than they might otherwise be, especially against moving tanks, but a range of payloads including standoff detonation and tandem warheads to defeat explosive reactive armour are available, increasing the weapon's ability to penetrate if a hit is scored.

In recent years, guided missiles of this sort have frequently been used as 'bunker busters' rather than anti-tank ordnance, and a weapon of this sort would certainly be useful in clearing defensive positions from the path of its dismounted troops.

Light tank variant

The IFV version is complemented by a light tank variant, built on the same hull and featuring the same swimming performance. This vehicle is fitted with a three-man turret mounting a 105mm (4.1in) rifled gun of NATO origins. Acting as a fire-support platform, the light tank variant carries high explosive and HE-frag rounds as well as armour-piercing fin-stabilized discarding sabot (APFSDS) anti-tank ammunition. Although its shells are not up to penetrating the latest generation of MBTs, the light tank can launch guided missiles through its gun.

As always for an amphibious vehicle, armour protection is limited. The ZBD-05 can withstand heavy machine gun rounds in the frontal arc and small-arms fire elsewhere, while some sources state that the front of the turret (on the tank variant) is armoured against 25mm (1in) armour-piercing rounds. In addition to the primary IFV and tank variants, an APC/armoured transport version exists. This may mount a 30mm (1.2in) cannon or a heavy machine gun.

Command, field ambulance and armoured recovery variants have also been produced, sharing the general performance and protection characteristics of the IFV version. The ZBD-2000 is in limited service with the Chinese military, equipping formations trained for the amphibious assault role. It does not appear to have been offered for export.

ZBD-2000 (ZTD-05)
Date 2006
Number Produced 86
Weight 28.2 tonnes (27.8 tons)
Length (with gun) 9.5m (31ft 2in)
Width 3.36m (11ft 1in)
Height 3.04m (10ft)
Engine Deutz diesel, 410kW (550hp)
Maximum Road Speed 65km/h (40mph)
Crew 4
Armour RHA
Main Armament Cloned 105mm (4.1in) ROF L7 L/52 rifled cannon
Main Gun Ammunition Stowed 46

Around 60 years after the great Sino–Soviet split, Chinese armoured vehicles such as this ZBD-05 have become a regular sight in joint exercises hosted by Russia.

ZBD-03

Date 2003

Number Produced Not Known

Weight 8 tonnes (7.9 tons)

Length 5.6m (18ft 4in)

Width 2.6m (8ft 6in)

Height 2.2m (6ft 5in)

Engine Diesel, 157kW (210hp)

Maximum Road Speed 68km/h (42mph)

Crew 3 (+ 5 infantrymen)

Armour Welded steel

Main Armament 1 x 30mm (1.2in) autocannon; 1 x HJ-73C ATGM

ZBD-03 (WZ506)

The ZBD-03 airborne combat vehicle was derived from the ZLC-2000. It has been designed for being air dropped by medium-sized transport aircraft such as the Ilyushin Il-76M.

SELF-PROPELLED GUNS

Early attempts to create self-propelled artillery were rather crude. In some cases the resulting vehicles were nothing more than a field artillery piece simply fixed atop a tracked chassis. Modern self-propelled guns are rather more sophisticated, and can in general be subdivided into anti-air weapons, indirect fire artillery platforms and direct fire-support vehicles that can function as tank destroyers.

This chapter includes the following self-propelled guns:

- Type 89 TD
- PTL-02 TD
- PGZ-88 SPAAG
- Type 80 SPAAG
- Type 95/PGZ-04 SPAAA
- PGZ-07/PGZ-09 SPAAG
- Yitian ADS
- Type 70-I
- Type 83
- PLZ-45
- PLZ-52
- PLZ-89
- PLZ-05
- PLL-09
- PLZ-07

The PLL-09 self-propelled artillery vehicle was developed by combining a PL96 122mm (4.8in) howitzer with the chassis of the ZBL-09 Snow Leopard APC.

Type 89

Some modern tank destroyers (or tank hunters as they are sometimes called) use guided missiles. Others mount a tank gun in a turret and are 'tank destroyers' mainly by virtue of the fact that they cannot fulfil the role of a battle tank due to a lack of armour protection.

The third class of tank destroyer is the expedient vehicle, created to plug a capability gap by mating a weapon system to an available chassis. The Type 87, or PTZ-89, falls into this category.

Cold War tank killer

During the 1970s, war with the Soviet Union seemed a possibility, and the new generation of Russian tanks led by the T-72 was of great concern to Chinese planners. A gun capable of penetrating the frontal armour of a modern MBT was sourced, in the form of a 120mm (4.7in) smoothbore of NATO specifications. However, an agreement to licence-produce the weapon could not be reached and a project to create a similar gun commenced.

After testing various designs and prototypes, an effective 120mm (4.7in) smoothbore was ready to be put into production. However, in the meantime, Chinese tank designers had selected a Russian-style 125mm (4.9in) weapon for use in the new generation of MBTs. Once in service, these tanks were expected to be an effective counter to their Russian equivalents, but in the interim a vehicle that could reliably penetrate Russian MBTs was needed. The solution was to create a tank destroyer out of the orphaned 120mm (4.7in) gun and the Type 321 armoured chassis.

Although the resulting vehicle had a turret and looked a lot like a tank, it was not capable of surviving if used as one. Weight and cost considerations required armour to be light – no more than 50mm (2in) at the front and less elsewhere – which meant that the Type 89 was destined to be a defensive sniper or ambush hunter.

The 120mm (4.7in) gun had a maximum range of 2500m (2734yd), with accuracy and penetration greatly reduced beyond 2000m (2187yd). A full range of ammunition was available, including high explosive and HE-frag, which might be used in an indirect fire-support role. The high elevation possible with the main gun gave a maximum range of 9km (5.5 miles) with these munitions. Tanks and armoured vehicles could be engaged with APFSDS and HEAT ammunition.

Type 89 Tank Destroyer
Date 1989
Number Produced 104
Weight 31.5 tonnes (31 tons)
Length (with gun) 5.6m (18ft 4in)
Width 2.8m (9ft 2in)
Height 3.12m (10ft 2in)
Engine 12V150LB/12150L liquid-cooled V-12-cylinder diesel, 388kW (520hp)
Maximum Road Speed 45km/h (28mph)
Crew 4
Armour (min–max) steel RHA 50mm (2in)
Main Armament 1 x 120mm (4.7in) L/50 smoothbore cannon
Main Gun Ammunition Stowed 30

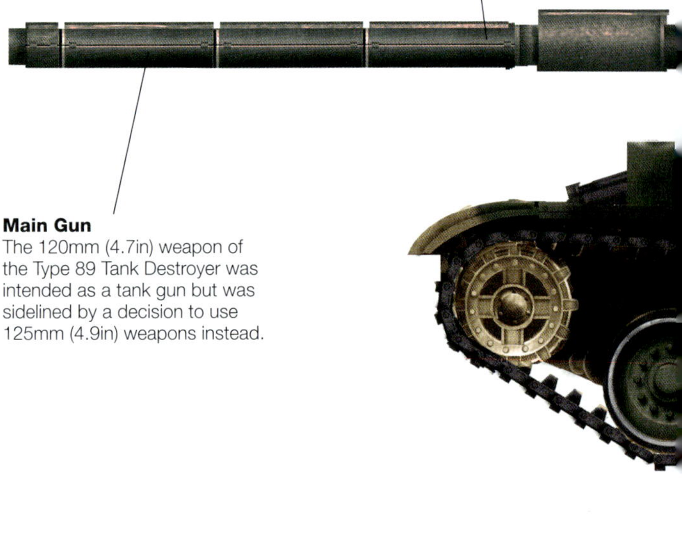

Range
The Type 89's gun can fire to a range of nine km (5.6 miles) at a rate of 10 rounds per minute.

Main Gun
The 120mm (4.7in) weapon of the Type 89 Tank Destroyer was intended as a tank gun but was sidelined by a decision to use 125mm (4.9in) weapons instead.

However, at any range where the Type 89 could hit an enemy tank, let alone have a decent chance of penetrating it, the converse was also true. The Type 89's gun was outranged by the Western weapons it was designed to emulate.

The Type 89 was apparently not well received by its users, and no more than 100 were produced in total. Although it had a laser rangefinder and some modern sighting aids, the gun was unstabilized. This might not be a problem when shooting from a stationary ambush position, but ideal conditions are rare in wartime. Consequently, Type 89s involved in a fluid engagement with almost any armoured vehicle, even one armed with a light autocannon, would be at a severe disadvantage.

In addition, the very light protection and relatively limited effective range made the vehicle highly vulnerable. It did not attract export customers and was retired from service in 2015.

PLA emblem and flag
The emblem of the People's Liberation Army is used across all branches of the Chinese armed forces, including the army, navy, air force, artillery and reserve units.

The flag of the PLA is red with a golden star in the upper left. The Chinese characters for 'Eight One' commemorate the establishment of the PLA during the Nanchang uprising on 1 August 1927.

Turret
Although the tank destoyer has a turret and appears like a tank, the gun is not stabilised so cannot fire accurately on the move.

PTL-02

The tank destroyer concept is all about creating a vehicle with the capability to engage heavily armoured vehicles – often with a tank gun – without the associated cost of a main battle tank.

PTL-02

The PTL-02 is designed for rapid deployment via airlift or under its own power, without requiring the specialist transports used for tracked vehicles.

Wheeled vehicles cannot carry the same level of protection as a tracked version, but can make effective anti-tank platforms – especially when armed with advanced weaponry taking advantage of improvements in targeting, propellants and recoil control.

The PTL-02 was based on the Type 92 APC, using the same 6x6 wheeled chassis. Other existing systems were also utilized, which helped keep development costs down. Among these was the main armament, a smoothbore 100mm (3.9in) gun developed from a towed anti-tank weapon. Its penetrator and explosive anti-tank munitions are effective against most armoured vehicles but could not take on a modern MBT with a realistic chance of success.

Classic tank destroyer

This might or might not be a serious drawback. Most of the targets a vehicle like the PTL-02 might encounter are not front-line main battle tanks fielded by a major power. Its weapon was sufficient for dealing with lighter vehicles and most older tanks, and the provision of high-explosive rounds allowed it to act as a fire-support platform. In the modern environment, a tank destroyer of this sort might be set against elderly first- or second-generation MBTs or even World War II-era tanks operated by a minor power, in which case it would be highly vulnerable to their weapons but quite capable of punching through their armour – the classic tank destroyer situation.

PTL-02 Tank Destroyer

Date 2001
Number Produced Not Known
Weight 19 tonnes (18.7 tons)
Length (with gun) 8.3m (27ft 3in)
Width 2.86m (9ft 5in)
Height 3.15m (10ft 4in)
Engine Deutz BF8L413F 8-cylinder turbo diesel 235kW (320hp)
Maximum Road Speed 85km/h (53mph)
Crew 4
Armour Not Known
Main Armament Type 86 100mm (3.9in) smoothbore gun
Main Gun Ammunition Stowed 30

PLL-05 Self-Propelled Gun-Mortar
Date 2008
Number Produced Not Known
Weight 16.5 tonnes (16 tons)
Length (with gun) 6.63m (21ft 9in)
Width 2.86m (9ft 5in)
Height 3.20m (10ft 9in)
Engine Deutz BF8L413F 8-cylinder turbo diesel 235kW (320hp)
Maximum Road Speed 85km/h (53mph)
Crew 4
Armour Not Known
Main Armament 120mm (4.7in) gun-mortar
Main Gun Ammunition Stowed 36

WMA-301 Assaulter

The WMA-301 Assaulter was developed for the international export market. It uses NATO-standard 105mm (4.1in) ammunition while most other parts are common with the Type 92 APC.

PLL-05

The PLL-05 is armed with a 120mm (4.7in) rifled gun-mortar derived from a very similar Russian weapon. It is capable of direct fire as well as indirect bombardment.

The PTL-02 would have an advantage over most older tanks in terms of accurate fire, as it was given sighting equipment developed from that of the Type 88 main battle tank. Combined with better mobility, it might well be able to pick off the tanks used by many minor powers without coming to harm.

Against a first-class opposing force it would fare badly, but could still be useful in a fire-support role.

Current service

The PTL-02 is in current service with the Chinese military, where it is used alongside vehicles based on a similar chassis. It has NBC protection and a tyre-pressure regulation system that permits it to cross soft ground, but has no amphibious capability. Waterjets or propellers and their associated machinery would increase the unit cost to create a capability that is not seen as necessary.

An export version designated WMA-301 has been taken up by a number of African nations as well as Myanmar. This variant has a distinctive wedge-shaped turret and mounts a 105mm (4.1in) gun compatible with standard NATO ammunition.

The International Army Games, held in Russia every summer since 2015, include a series of mobility and shooting contests for tanks, IFVs and fire-support vehicles such as the PLL-05.

PGZ-88 SPAAG

The concept of the self-propelled anti-aircraft weapon originated before World War II but became essential during that conflict, at a time when short-range attack by relatively slow-moving aircraft was a constant threat.

PGZ-88 SPAAG

The PGZ-88 attempted to keep costs down by utilizing an existing tank chassis, but ended up creating a vehicle that was too heavy and at the same time ineffective.

Mounting a heavy machine gun on the turret of a tank or the cab of a truck created a measure of defence, but it was obvious that more concentrated firepower would be necessary.

The earliest true self-propelled anti-aircraft guns (SPAAG) were created by mating twin- or quad-mounted machine guns or light cannon to a mobile chassis. This was often a half-track truck or sometimes a fully tracked chassis, and it was often the fate of obsolete tank designs to be converted into platforms for anti-aircraft weapons. These were naturally incorporated into the armoured formations to provide protection against air attack.

The extensive use of helicopters in early Cold War conflicts such as the Vietnam War added an extra dimension to the air defence equation. Multiple machine guns were no longer effective, but cannon in the 20–35mm (0.8–1.4in) range provided a balance of hitting power and volume of fire that could bring down helicopters or fast-moving attack jets.

US pilots developed 'rules for survival' stipulating the altitudes necessary to counter different threats, but the development of effective surface-to-air missile (SAM) systems made the situation a lot more complex. Early SAMs could not engage a low-flying target, so a general rule for pilots was to stay above the ceiling for light weapons fire unless SAM emplacements were known to be in the area or a launch was detected.

PGZ-88 SPAAG

Date 1989
Number Produced Not Known
Weight 35 tonnes (34.4 tons)
Length (with gun) 6.24m (20ft 6in) (based on Type 79 chassis)
Width 3.31m (10ft 10in) (based on Type 79 chassis)
Height Not Known
Engine Diesel 425kW (570hp)
Maximum Road Speed 50km/h (31mph)
Crew 4
Armour Not Known
Main Armament 2 x 37mm (1.6in) cannon
Main Gun Ammunition Stowed 1000 rounds per gun

The appearance of large-calibre autocannon in fixed emplacements and carried aboard vehicles made low flight to avoid the missiles dangerous. The combination of both could be lethal, with the threat of one forcing aircraft into the killing zone of the other. Fixed weapons could be plotted and perhaps avoided, but heavy self-propelled anti-aircraft systems made it possible to stage an ambush by creating a concentration of anti-aircraft guns and using a SAM launch to force enemy aircraft down into it.

Radar guidance

Many early SPAAGs were manually aimed and thus largely ineffective, although volume of fire could still achieve hits. Radar guidance greatly improved lethality and was implemented as soon as a set small enough to be carried on a vehicle turret became available. The Russian ZSU-57, entering service in 1955, and the quad-barrelled ZSU-23 that appeared in 1962, demonstrated what could be done with multiple cannon on a tracked chassis. For their first attempt at a similar vehicle, Chinese designers selected a twin 37mm (1.6in) cannon mount capable of elevating to 85 degrees.

The guns, developed from a similar Russian weapon by way of a towed anti-aircraft system, are guided by a search radar with a range of around 15km (9 miles) and targeting radar plus a ballistic computer, enabling engagement of airborne targets out to more than 7000m (7655yd). Their high-explosive shells are effective against most unarmoured targets, and the high elevation of anti-aircraft weapons has at times proven useful when engaging non-aircraft targets. When operating in very mountainous terrain or urban environments, vehicles of this type can engage targets beyond the elevation of weapons carried by standard IFVs or tanks. Thus SPAAG vehicles can sometimes be encountered operating where there is little or no air threat. The guns were combined with a Type 79 tank chassis, which did away with the need to develop a vehicle specifically to carry the weapons. The resulting mobile anti-aircraft system was rather heavy for its role at 35.5 tonnes (35 tons), compared to the 19-tonne (19-ton) Russian ZSU-23. It is likely the armour of the chassis was unchanged from the original tank design, creating an over-protected hull that did not prevent the vehicle from being 'mission killed' by a hit on its lightly armoured turret or exposed radar.

Failed type

The PGZ-88, sometimes referred to as Type 88, was not accepted for service by the Chinese military. Its combination of excess weight and small-arms-only protection for the armament and weapon crew resulted in rejection after trials. If it was ever offered for export, it did not attract any buyers.

Type 63 SPAAG

Combining the chassis of a T-34/Type 58 medium tank and an open turret with a pair of 37mm (1.5in) anti-aircraft guns, this 'lash-up' was considered mediocre at best. The Type 63 was supplied by China to the North Vietnamese Army (NVA) during the Vietnam War. It remained in service with the post-war People's Army of Vietnam, as well as the PLA, with the last examples being retired in 1990.

Type 63 SPAAG
Date 1968
Number Produced Not Known
Weight Not Known
Length (with gun) 5.9m (19ft 7in)
Width 3m (9ft 10in)
Height 2.6m (8ft 7in)
Engine V-2 V12 diesel 372kW (493hp)
Maximum Road Speed 50km/h (31mph)
Crew 6
Armour Not Known
Main Armament 2 x 37mm (1.45in) Type 63 AA guns

Type 80

The next attempt at a mobile anti-aircraft platform was a near-copy of the Russian ZSU-52-2. This system dated from 1955 and was obsolete by the 1980s, but Chinese designers created a very similar vehicle nonetheless.

Type 80 SPAAG

Although impressive in appearance and on-paper capabilities, the twin-57mm (2.2in) self-propelled anti-aircraft platform was not any more effective in Chinese service than with any other nation.

Type 80 SPAAG
Date 1980s
Number Produced Not Known
Weight 29 tonnes (28.54 tons)
Length (with gun) 8.5m (27ft 10.5in)
Width 3.27m (10ft 8.5in)
Height 2.75m (9ft)
Engine 12150L-7BW V12 diesel 432.5kW (580hp)
Maximum Road Speed 55km/h (34mph)
Crew 6
Armour Not Known
Main Armament 2 x 57mm (2.2in) cannon
Main Gun Ammunition Stowed Not Known

At its heart was a pair of 57mm (2.2in) cannon derived from a Russian design intended to arm divisional air defence formations during the 1950s.

Like its predecessor, the PGZ-88, the Type 80 was mounted on the chassis of a Type 79 (or type 69III-A) main battle tank. Some sources state that the chassis was given reduced armour protection: the Type 80 was certainly significantly lighter than the PGZ-88, at 29.4–31.5 tonnes (29–31 tons) instead of 35.5 tonnes (35 tons). However, it shared many of the same deficiencies.

Excess weight

The Type 80 was far heavier than it needed to be – the Russian twin 57mm (2.2in) mobile AA system used a smaller and lighter hull with four roadwheels to the Type 80's five. This large hull was armoured to a level that an anti-aircraft system would be unlikely to ever need, which in turn required a large and fuel-hungry engine to move it around. This all added up to a lack of cost-effectiveness both in terms of the initial purchase and during the vehicle's service life. Little benefit was gained from the armoured chassis, since the large turret was more likely to be hit by most attacks the vehicle might face, and disabling its weapons would put it out of action just as surely as a penetrating hit on the hull.

The Type 80 is claimed to have had an effective range of around 4000m (4374yd) using manual sighting and 6000m (6561yd) with radar. It has seen service, as have similar weapons produced by other countries, in several conflicts but has never proven very effective. Many former users have converted their vehicles to other roles or removed them from service.

A twin 57mm (2.2in) cannon mount on an armoured chassis does have potential in defensive and fire-support applications. In a world where many potential opponents have no combat vehicles tougher than a civilian pickup truck fitted with a heavy machine gun, surviving Type 80s and similar platforms might be repurposed as ground combat assets. The Type 80 is still inefficient and unevenly protected in this role, but it could be useful. As an air defence platform, however, it was marginal when it was introduced and has rightfully been supplanted by better vehicles.

Type 95/PGZ-04

Combination gun/missile air defence systems have been attempted by various nations. In theory they offer many possibilities, with missiles engaging fast-moving targets and guns making the local area untenable for helicopters. However, such systems tend to be plagued with difficulties and rarely perform as well as their designers hope.

Type 95 SPAAA

Quad 25mm (1in) cannon offer a good compromise between the volume of fire necessary to make hits likely and the ability to inflict significant damage.

The Type 95 can operate solo but is designed to be deployed as part of an integrated unit containing one command and six combat vehicles all sharing a common chassis, with additional supporting vehicles of other designs. The Type 95 weighs in at 22.8 tonnes (22.5 tons). This is much lighter than preceding designs and suggests a more sensible balance between hull protection and capability.

Soviet copy

The gun component of the Type 95 is apparently inspired by the Soviet ZSU-23-4 '*Shilka*', but uses quad-mounted 25mm (1in) cannon instead of the *Shilka*'s 23mm (0.9in) weapons. It is also possible that the design was influenced by the Italian SIDAM 25 self-propelled air defence system, which was examined by Chinese experts in the late 1980s. SIDAM 25 used a quad 25mm (1in) mount on a modified M113 APC chassis. The vehicle as a whole was not exported, but the turret and weapon systems have been mounted on other vehicles by overseas buyers, proving the viability of a quad 25mm (1in) AA system.

The guns on the Type 95 are of 1980s vintage, having first been deployed as a towed air defence weapon. They are directed by a low-

Type 95/PGZ-04 SPAAA

Date 1999
Number Produced Not Known
Weight 22.86 tonnes (22.5 tons)
Length 6.71m (22ft)
Width 3.2m (10ft 6in)
Height 3.4m (11ft 2in) (radar down)
Engine Not known
Maximum Road Speed 53km/h (33mph)
Crew 3
Main Armament 4 x 25mm (1in) cannon; MANPAD missiles
Range Guns: 2500km (2734yd); missiles: 6000m (6562yd)
Rate of Fire Gun: 3200rpm

This Type 95 SPAAG is on display in Beijing's PLA Tank Museum. Highly mobile gun/missile air defence systems offer tank formations a defence against their natural predator, the attack helicopter.

altitude radar system carried aboard the Type 95, or can receive targeting data from the command vehicle. Electronic assistance allows reasonably accurate fire even when the vacuole is moving, against ground targets as well as aircraft. This gives air defence vehicles a secondary fire-support capability and makes them useful escorts for convoys moving through an area suspected to be infiltrated by insurgents.

The guns are supplemented by four QW-2 missiles. The QW-2 is a short-ranged weapon developed as a shoulder-fired man-portable air defence system (MANPADS). Early thermal-guided weapons of this type were very limited and could only track a target with a very strong heat signature, and then only from angles that did not mask the hottest parts of the aircraft. Third-generation missiles such as the QW-2 are described as 'all-aspect' weapons, capable of tracking an aircraft coming directly towards the sensor or from any other angle.

Short-range weapon

Other improvements since the first man-portable heat-seeking missiles were deployed include a greater resistance to being distracted by defensive flares, but this is still a short-ranged weapon. It might be effective against fast jets under the right circumstances, but a low-flying aircraft is likely to cross the weapon's engagement envelope before it can react, and one flying above 3500m (3827yd) cannot be engaged at all. However, light missiles of this sort are effective against helicopters, enabling the Type 95 to protect armoured formations from tank-hunting gunships.

The vehicle's radar can detect targets out to about 11km (7 miles) and track them within 6km (4 miles). Reaction time from detection to firing is given by Chinese sources as around 10 seconds using electronic sighting, and about half that with optical sighting. The Type 95 is in service with the Chinese military but does not seem to have generated overseas interest.

Type 95 SPAAA

The high elevation possible with air defence weapons makes them useful for engaging other threats, such as RPG teams and snipers located high in buildings or the sides of a mountain pass.

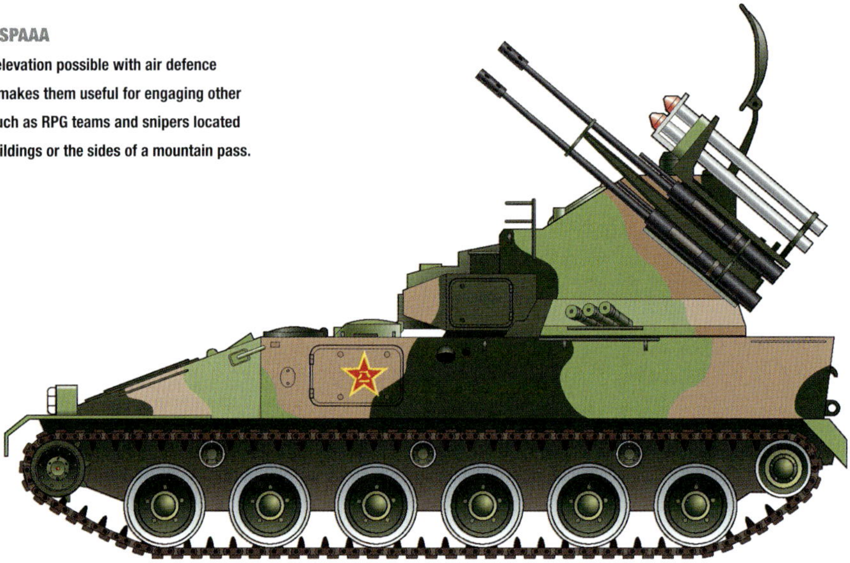

PGZ-07/PGZ-09 SPAAG

Originally designated PGZ-07 and apparently redesignated PGZ-09, this vehicle mounts a pair of 35mm (1.4in) cannon and may be able to carry additional short-ranged missiles.

It is constructed on the same chassis (or a near-identical one) as the Type 95 self-propelled anti-aircraft system, with the engine at front-right and a centrally mounted turret.

Basic protection

Performance is suitable to keep pace with an armoured formation, with a road speed of 55km/h (34 miles) and an operating radius of around 450km (280 miles). Protection is sufficient to prevent penetration by small-arms fire, but the vehicle is vulnerable to even heavy machine guns. Despite this, it is intended to serve in a fire-support role as a secondary capability.

In addition to high-explosive-incendiary (HEI) and HEI-tracer rounds, the guns have a semi-armour-piercing explosive round available that poses a severe threat to lightly armoured vehicles or infantry positions.

Tracking and engagement of targets is by means of two radar sets and an electro-optical sighting system. Surveillance and detection radar is located at the rear of the turret, with tracking radar and other sensors at the front. It is not certain what systems are carried, but it is likely that the electro-optical fit includes thermal and low-light cameras as well as a laser rangefinder.

Sighting system

The range of airborne targets to be engaged by air defence vehicles continues to expand. Fast jets pose a different tracking and targeting problem to helicopters, but air defence units

PGZ-07/PGZ-09 Self-Propelled Anti-Aircraft Gun

Date 1999
Number Produced 1000
Weight Not Known
Length (with gun) 6.7m (22ft)
Width 3.2m (10ft 6in)
Height Not Known
Engine Not Known
Maximum Road Speed 55km/h (34mph)
Crew 3
Armour Not Known
Main Armament 2 x 35mm (1.34in) cannon
Main Gun Ammunition Stowed Not Known

PGZ-09

The electro-optical sighting system of this PGZ-09 is clearly visible above the tracking radar at the front of the turret.

PGZ-07

It is not clear what changes were made, if any, between the PGZ-07 pictured here and the current PGZ-09.

may also have to deal with drones and missiles. Even quite large drones tend to have a small thermal and radar signature, while missiles are small, fast, and may be flying an evasive course close to the ground.

Some of the engagement problems are mechanical, such as slewing the turret fast enough to get the guns on target and keeping them there, while others require increasingly sophisticated computers to calculate an intercept between shells and a small, fast-moving target.

Modern air defence vehicles thus need to be extremely sophisticated and are therefore expensive to develop and upgrade. China's first attempts in this arena were not very promising, but the only ways to develop a mature capability are to make mistakes or to buy vehicles from those who have made their own – at whatever price they demand. The latter approach was clearly unacceptable to the Chinese military, leaving an expensive development process as the only option.

Yitian Air Defence System

The Yitian air defence system is built around the TY-90 (Tian Yan) missile. This project was initially designed to give helicopters a measure of air-to-air capability, which is of interest to several nations, and was later converted for use on ground-based launchers.

The missiles use thermal guidance, and may have a dual-mode capability that makes them greatly more resistant to flares dropped as decoys. The Yitian system's radar can detect and track a strike aircraft at 18km (11 miles), and a large missile at 8km (5 miles). Engagement range is around 6km (4 miles).

Standoff and Cruise missiles are an ever-increasing threat in the modern combat environment, with many nations either possessing missiles obtained from major manufacturers

Yitian Air Defence System (WMZ551)
Date 2004
Number Produced Not Known
Weight 16 tonnes (15.74 tons)
Length 6.63m (21ft 9in)
Width 2.8m (9ft 2in)
Height 2.89m (9ft 6in)
Engine Deutz BF8L413F 8-cylinder turbo diesel 235kW (320hp)
Maximum Road Speed 85km/h (53mph)
Crew 3
Armour Not Known
Main Armament 8 x TY-90 missiles
Main Gun Ammunition Stowed No reloads carried

Ytian ADS

Guidance for the Yitian's eight short-range air defence missiles is provided by a radar mounted above the launchers, creating a self-contained module that can be fitted to a variety of vehicles.

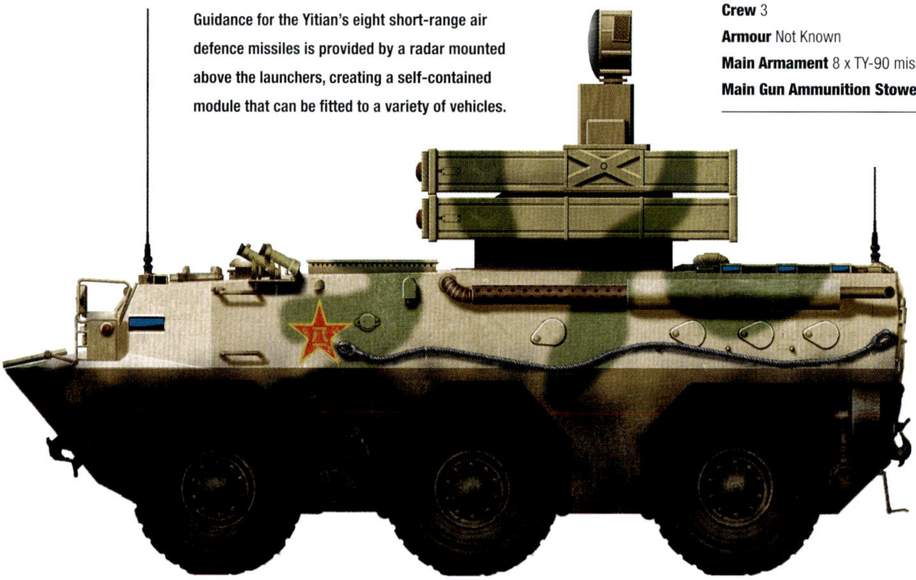

or producing their own. The more advanced missile systems are capable of navigating over very long distances and making an attack so precise that one missile can fly though a hole made by another to explode inside a structure.

Missiles with various types of warhead can be launched against installations and infrastructure as well as military formations, and a mobile short-range interception capability is increasingly a necessity. Vehicle-mounted air and missile defence units can be deployed to protect dams, power stations and other assets that might be targeted as part of a softening-up process before invasion, and can be transferred between locations or assigned to protect combat units once the threat to infrastructure recedes.

The Yitian air defence system has been mounted on a variety of vehicles, notably the W321 6x6 wheeled APC chassis. A relatively light weapon system of this sort does not really need a tracked chassis, and a missile-armed air defence unit has no business going where anti-armour weapons will be fired at it by ground troops. Thus a wheeled version makes sense, reducing costs and easing maintenance while allowing rapid redeployment under its own power.

APC mount

The W321-based Yitian system is equipped with two sets of quad-mounted missiles along with the usual machine gun for self-defence. Tracking and guidance radar is mounted above the missiles, with an electro-optical sighting system also available to the crew. The version of the W321 chassis used is probably the same as the Type 92 family of vehicles, though some modifications may have been necessary in order to accommodate the missile mount.

HonQi HQ7B

The HonQi 7B missile system was reverse-engineered from the French Crotale. It is deployed aboard a variety of light vehicles as well as on towed mounts. In this case there are four missiles. Each one is 3m (9ft 9in) long and weighs 84.5kg (186lb). The missiles have a range of 15km (9.3 miles).

Type 70-I

The Type 70-I is a turretless self-propelled howitzer using a weapon based on the 122mm (4.8in) Soviet M-30. This rather elderly artillery system was developed in the 1930s and produced into the mid-1950s. Widely exported, the M-30 and its derivatives have seen combat in numerous conflicts where it has proven simple but effective.

Before the Sino–Soviet split of the 1950s, a programme to build a local copy of the M-30 was set up with Russian assistance. These guns were available in large numbers and, designated Type 54, served the Chinese military as a towed support weapon. Although as effective as any other gun once in position, the Type 54 was vulnerable to counter-battery fire and could not keep pace with an armoured advance or counterattack. Self-propelled artillery that could 'shoot and scoot' was an obvious necessity, but creating a suitable vehicle would take time.

In the interim, a mobile artillery platform was created using the YW531H APC – a modified version of the Type 63 APC chassis – as a basis. Slightly longer than a standard Type 63, the vehicle that emerged was an obvious interim solution, mounting the Type 54 gun – shield and all – in a turretless configuration atop the hull. This created an artillery piece with tracked-vehicle mobility and provided protection to the crew on the move, but left them vulnerable when operating the weapon.

Around 200 of these vehicles, designated Type 70-I, were produced before true self-propelled artillery became available. The Type 70-I

Type 70-I

A turretless howitzer was simple to produce and under most circumstances quite adequate, but left the crew vulnerable when operating in a direct-fire role.

Type 70-I
Date 1970
Number Produced 200
Weight 15.4 tonnes (15 tons)
Length 5.65m (18ft 6in)
Width 3.06m (10ft)
Height 2.68m (8ft 9in)
Engine Deutz 6150L 6-cylinder diesel, 192kW (257hp)
Maximum Road Speed 56km/h (35mph)
Crew 7
Armour Not Known
Main Armament Type 54 122mm (4.8in) howitzer
Main Gun Ammunition Stowed 40

entered service in 1970 and was followed by an improved version designated Type 85. Some sources state that the Type 85 was an export version and did not enter Chinese service. The Type 70 itself did not achieve export success.

In addition to leaving the gun and crew unprotected, the Type 70-I retained all the shortcomings of the Type 54 towed howitzer. China's first indigenously produced large-calibre artillery weapon, it was capable of firing a 122mm (4.8in) shell out to 12km (7 miles) or so but lacked modern sighting systems and was inaccurate. This might or might not be a problem for artillery bombarding static enemy positions but in the fluid environment of an armoured engagement the lack of precision was a serious drawback.

The Type 54 howitzer atop the Type 70-I normally delivered high-explosive shells but was also capable of firing a range of munitions, including smoke and incendiary ammunition. A shell containing HEAT submunitions has also been developed, intended to attack the thinner top armour of enemy vehicles by bursting overhead. Given the inaccuracy of the Type 70-I's weapon, the effectiveness of this ammunition remains debatable.

Assault gun role

Enclosed self-propelled artillery has been used in the assault gun role on numerous occasions, with a good success record. The high angle of gun elevation allows return fire at targets a tank could not engage, and this was used by Israeli forces operating in Beirut in the early 1980s to deter snipers operating from the upper floors of buildings. The Type 70-I was capable of conducting direct fire in this manner or against ground-level targets, but without crew protection this would have led to significant casualties.

The Type 70-I was, overall, a quick-and-dirty stopgap vehicle intended to provide mobile artillery support at a time when none existed. Although crude by modern standards it was an enormous improvement on towed artillery, which might be as much of an encumbrance as an assistance to fast-moving armoured forces. It was quickly surpassed by more advanced mobile artillery systems and was not adopted by overseas armed forces.

Type 85
Date 1985
Number Produced 100
Weight 18.2 tonnes (18 tons)
Length 6.66m (21ft 9in)
Width 3.06m (10ft)
Height 2.85m (9ft 4in)
Engine Deutz Type BF8L413F diesel, 238kW (320hp)
Maximum Road Speed 60km/h (37mph)
Crew 6
Armour Not Known
Main Armament Type 85 122mm (4.8in) howitzer
Main Gun Ammunition Stowed 40

Type 85 (YW323)

The longer gun of the Type 85 offered range and accuracy advantages, but this was still a rather basic vehicle which was quickly replaced by better systems.

Type 83

The Type 83 self-propelled gun is built around the Chinese PL-66 howitzer, which in turn was derived from the Russian D-20.

Using 152mm (6in) ammunition, this is one of the most widely used artillery pieces in the modern world and was the basis of the Russian 2S3 Akatsiya self-propelled gun. The PL-66 is rather dated now, though still capable of being effective on the battlefield, but remains in service with the Chinese military.

In its guise as the self-propelled Type 83, the PL-66 has a 27-calibre barrel. This is a little longer than the towed version but short by modern self-propelled gun standards. Although this does simplify transportation and movement through cluttered terrain, it also means the Type 83 has a relatively short effective range. With conventional shells this is around 17km (10 miles), although a rocket-assisted projectile is available with an extended range of around 22km (14 miles).

The Type 83 was heavily influenced by the Russian 2S3 Akatsiya, which is capable of firing laser-guided shells. China acquired this technology, so it is probable that the Type 83 has a similar capability. Laser-guided munitions can be used against any precision target, notably bunkers and fortified positions, but are typically anti-tank weapons. Laser guidance is not infallible, as it requires the designator to maintain line-of-sight on the target, but it does allow distant artillery units to attack moving tanks without being exposed to an immediate response.

The Type 83 was built on the same W321 chassis as many other Chinese armoured vehicles. Its 410kW (550hp) engine allows a maximum speed of 55km/h (34mph) and an operating radius of 450km (280 miles) – sufficient to keep up with armoured or mechanized forces. Armour protection is sufficient to stop small arms or shell splinters – the latter being the most likely threat if counter-battery fire finds the vehicle before it can change firing positions – and a heavy machine gun is provided for self-defence.

Type 83

The Chinese military has selected 122mm (4.8in) guns for its divisional artillery. Larger-calibre weapons like the Type 83 are assigned to specialist formations.

Type 83 Self-Propelled Howitzer

Date 1983
Number Produced Not Known
Weight 30 tonnes (29.5 tons)
Length (with gun) 7.33m (24ft 0.5in)
Width 3.23m (10ft 7in)
Height 3.5m (11ft 6in)
Engine 12150L diesel 410kW (550hp)
Maximum Road Speed 55km/h (34mph)
Crew 4–5
Armour Not Known
Main Armament 1 x 152mm (6in) howitzer
Main Gun Ammunition Stowed 30 Rounds

PHZ-89 SP MRL

Using its self-reloading system the PHZ-89 can deliver two rapid salvoes of 40 122mm (4.8in) rockets and be quickly on the move to avoid counter-battery fire.

Multiple rocket system

The same chassis is used as the basis for a multiple rocket system designated Type 89 or PHZ-89, and has similar performance.

The PHZ-89 rocket system uses 40 launching tubes for 122mm (4.8in) artillery rockets and carries a complete set of reloads in a covered rack at the front of the hull, permitting two salvoes to be launched in rapid succession. Reloading time is around two minutes from the ready rack, and considerably longer if manually reloading the vehicle.

PLZ-45 (Type 88)

China started the Cold War arms race far behind the Western powers and the Soviet Union, and needed to catch up quickly. Once political differences ended Soviet assistance, the usual expedient was to develop local versions of foreign equipment.

This usually led to Chinese hardware being a decade or more behind the other major powers, but at times the process of catching up was short-circuited by obtaining access to the most advanced weaponry on the market. The Austrian GHN-45 howitzer was one such weapon. From it, Chinese designers developed a 155mm (6.1in) towed artillery weapon that entered

The PLZ-45 was designed for the export market, and most examples serve overseas. A small number are retained by the Beijing military district.

service in 1991 under the designation PLL-01 or Type 89. This represented a move from Soviet calibres to NATO standard ammunition, which would eventually make Chinese military equipment more attractive to overseas buyers. In the meantime, their armed forces got a modern artillery piece and advanced ammunition for it.

Along with the GHN-45 gun, China gained the technology to produce extended-range artillery ammunition that used aerodynamics rather than rocket boosting. Continued development has reportedly produced shells with a range double that of standard 155m (6.1in) artillery ammunition. The PLL-01 towed howitzer can also deliver precision-guided shells using laser designation, and has a semi-automatic loading system.

PLL-01 gun

The PLZ-45 is a self-propelled artillery system built around a version of the PLL-01 gun and loader. Unlike some previous Chinese armoured vehicles, an existing chassis was not used. The PLZ-45 was designed from the ground up to carry its weapon system, and was complemented by an ammunition carrier/loader vehicle built on the same chassis. It is capable of operating in the direct-fire role, but as the vehicle is only protected against small arms and shell splinters this would be risky.

When operating in a more typical indirect support mode, the PLZ-45 uses GPS positioning and an automatic gun-laying system to enhance accuracy even when not firing laser-guided projectiles. The latter have a range of about 20km (12 miles), whereas extended-range unguided projectiles include high explosive, HE-frag and smoke as well as cargo shells capable of delivering a range of payloads to the target. Projectiles are loaded automatically at any angle of elevation, with propellant loaded manually by the vehicle's two loaders.

The PLZ-45 was offered for export and taken into service with Algeria, Kuwait and Saudi Arabia. It also serves in very limited numbers with the Chinese armed forces, although the PLZ-05 was adopted for widespread service instead.

PLZ-45 Self-Propelled Gun
Date 1997
Number Produced Not Known
Weight 32 tonnes (31.5 tons)
Length 6.1m (20ft)
Width 3.2m (10ft 6in)
Height 2.59m (8ft 6in)
Engine Diesel, 391.4kW (525hp)
Maximum Road Speed 56km/h (35mph)
Crew 5
Main Armament 1 x 155mm (6.1in) WAC-21 howitzer
Main Gun Ammunition Stowed 30

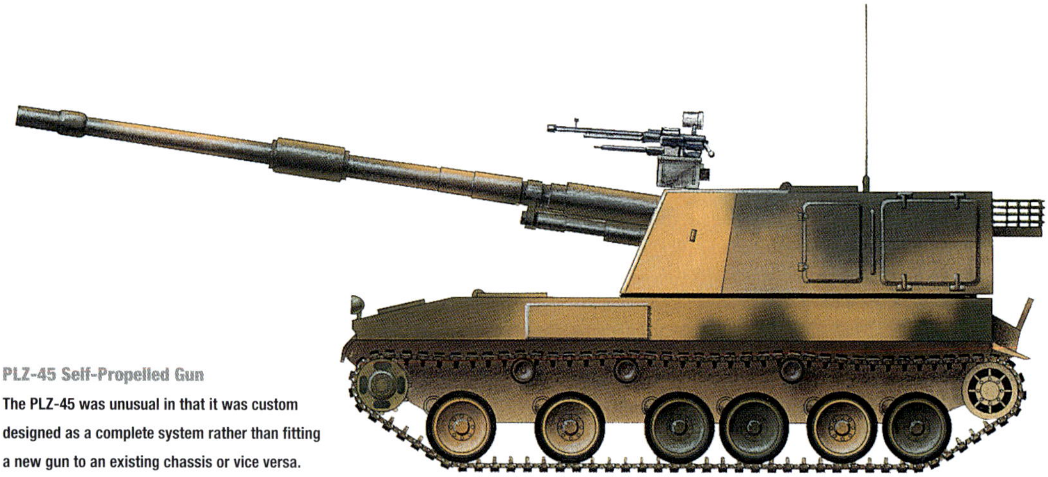

PLZ-45 Self-Propelled Gun
The PLZ-45 was unusual in that it was custom designed as a complete system rather than fitting a new gun to an existing chassis or vice versa.

PLZ-89

The PLZ-89 (or Type 89) bears a distinct resemblance to the Russian 2S1 *Gvozdika*, which entered service in 1971.

The latter uses a gun derived from the Russian 122mm (4.8in) D-30 towed howitzer, while the Type 89 is built around the Chinese Type 86 (or W86), which is more or less a direct copy. This weapon was taken into Chinese service in the 1980s and offered for sale on the export market as an alternative to the Russian weapon.

Fire support

The standard level of artillery support was set at 122mm (4.8in) in the late 1980s, so a suitable platform to carry such a weapon was needed. Several designs for self-propelled artillery vehicles were trialled until eventually what would become the Type 89 was adopted. This was a conventional self-

PLZ-89 Self-Propelled Howitzer

The PLZ-89 has a distinctly Russian look about it, which is not surprising considering the lineage of its main gun.

PLZ-89 Self-Propelled Howitzer

Date 1989
Number Produced Not Known
Weight 20 tonnes (19.68 tons)
Length (with gun) 6.5m (21ft 4in)
Width 3.4m (11ft 2in)
Height 2m (6ft 7in)
Engine 12V150L12 Diesel 336kW (450hp)
Maximum Road Speed 60km/h (37mph)
Crew 5
Armour Not Known
Main Armament 122mm (4.8in) howitzer
Main Gun Ammunition Stowed 40 Rounds

Battlegroup organization

As the Chinese armed forces became increasingly mechanized, a shift in organization and doctrine occurred. Today's Chinese forces are modular in nature, allowing a battlegroup of the appropriate size to be created by a parent division. Armoured and infantry assets are assigned as needed and given support from the division's pool of artillery and air defence units.

This model is becoming increasingly common worldwide. An armoured or mechanized infantry division is in theory an extremely potent force but what matters is the force brought to bear at the critical point. A traditional brigade or division may not be best suited to do so. A battlegroup system allows a tailored force to be assigned to a given combat area, concentrating the division's assets where they can be the most effective and providing support as necessary. This is particularly useful when engaging forces that are not neatly organized into conventional brigades and divisions, or in a situation where a division must cover a wide area where the intensity of conflict can vary considerably. To enable this structure to function, vehicles must have roughly equivalent performance.

propelled howitzer armoured against small-arms fire and mounting a heavy machine gun for self-defence. Its chassis was developed from the Type 77 amphibious APC, but it was not constructed for amphibious capability. With the addition of a flotation kit, the Type 89 was able to cross calm waters.

SH-3

The SH-3, entering service in 2007, is an improved version of the Type 89 with a more powerful engine. It has a semi-automatic loader and improved fire control, enabling it to deliver more rounds on target than its predecessor. It has been offered for export but sales have thus far been very limited.

PLZ-52

The PLZ-52 is visually similar to the PLZ-45 but is larger and has a much more powerful engine. It was first publicly unveiled in 2014 and is intended for the export market.

The PLZ-52's 155mm (6.1in) gun is longer than that of the PLZ-45, generating a higher muzzle velocity that increases range and accuracy. It is compatible with standard NATO ammunition and is fed by a modular charge system.

One advantage of self-propelled artillery is the ability to 'shoot and scoot', moving on quickly to avoid counter-battery fire. It is by no means certain that artillery will be targeted in this way, but as soon as a weapon opens fire its position can be determined by tracking shell trajectories using radar. Thus, the clock starts ticking once the first shot is fired.

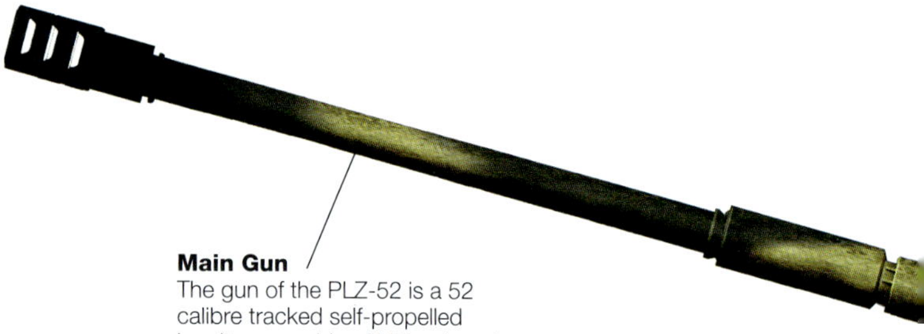

Main Gun
The gun of the PLZ-52 is a 52 calibre tracked self-propelled howitzer capable of hitting targets at a range of 53km (33 miles) with ERFB-BB-RA projectiles. It can fire up to eight rounds per minute.

Precision-guided munitions

Computerized gun laying eliminates the need for ranging shots, which not only delays an enemy response but also increases the chance that the target will be caught by surprise in the open. The ability of a battery to get the requisite number of shells on target and quickly be ready to move greatly improves survivability, and reloading time is an important component of this equation. Thus, loading systems are

PLZ-52 Self-Propelled Gun
Date 2014
Number Produced Around 250
Weight 43 tonnes (42.3 tons)
Length (with gun) 11.6m (38ft 0.5in)
Width 3.38m (11ft 1in)
Height 3.55m (11ft 8in)
Engine Diesel 745kW (1000hp)
Maximum Road Speed 65km/h (40mph)
Crew 4
Armour Not Known
Main Armament 155mm (6.1in) howitzer
Main Gun Ammunition Stowed 30–40

not merely labour-saving devices or enhancements to the unit's firepower; they reduce the period in which a battery is vulnerable to counter-fire.

In addition to precision-guided munitions and extended-range shells with various warheads, the PLZ-52's gun can deliver GPS-guided shells or an anti-armour payload consisting of two winged submunitions. The cargo shell probably has a range of 30–35km (19–22 miles), delivering 'smart' submunitions to the target area. These then seek a suitable armoured target and guide themselves to it using their winglets.

Munitions intended to allow artillery to destroy tanks at a great distance have been in development since the 1960s, but have always posed serious technical challenges. It may be that an effective armour-seeking artillery weapon will change the nature of armoured engagements. Tests in recent years have produced very promising results, but these weapons are yet to be tested on a battlefield and their long-term influence remains debatable.

Protection

The PLZ-52 is armoured against shell splinters and small-arms fire, and has an NBC protection system. Vehicles offered for export may not include this feature, which is deemed unnecessary by many potential clients and drives up the purchase cost considerably. The PLZ-52 itself carries 30–40 ready rounds, but these can be used up very quickly. It is complemented by a resupply vehicle built on the same chassis. This is reportedly capable of carrying 90 rounds of 155mm (6.1in) ammunition and delivering them to the gun at a rate of eight rounds per minute using its own crane and transfer mechanism.

PLZ-52 Self-Propelled Gun

The PLZ-52 can deliver tank-seeking munitions to a distant target area, but whether these weapons will make heavy armoured vehicles obsolete remains to be seen.

PLZ-05

Entering service in 2008, the PLZ-05 replaced the Type 83 self-propelled gun. This marked a transition from Russian to NATO calibres as well as an increase in firepower, which was augmented by rapid-loading systems that allow a battery to have several shells in the air at once, and computerized gun-laying systems that can time a salvo to arrive all at once.

The PLZ-05 was developed from the PLZ-45 and is also built around a 155mm (6.1in) gun. This weapon requires a large turret, both to house the loading and firing mechanism and to permit a sufficient amount of ready ammunition to be carried. The PLZ-05's gun is longer than that of its predecessors – 52 calibres rather than 45 – which gives greater range and accuracy.

The PLZ-05 has not been offered for export, but serves with the Chinese armoured forces as well as some coastal artillery formations. The idea of coast defence guns may seem outdated in an age of guided missiles, but modern artillery can reach 30km (19 miles) or more out to sea, making it extremely hazardous to approach the coast, and poses a severe threat

Main Gun
The gun of the PLZ-05 is a 52-calibre weapon, i.e. its length is 52 times the diameter of its shell. This is significantly longer than the 45-calibre gun of the preceding PLZ-45.

Powerplant
The 745kw (1000hp) engine is located at the front right of the vehicle, driving through front sprockets. The tracks are guided by three return rollers.

PLZ-05 Self-Propelled Gun
Date 2005
Number Produced 200+
Weight 35 tonnes (34.4 tons)
Length (with gun) 11.6m (38ft 1in)
Width 3.4m (11ft 2in)
Height 3.55m (11ft 8in)
Engine 8V50 turbo diesel 600kW (800hp)
Maximum Road Speed 65km/h (40mph)
Crew 4
Armour Not Known
Main Armament 155mm (6.1in) howitzer
Main Gun Ammunition Stowed 30–40

PLZ-05 Self-Propelled Gun

The main gun can fire up to eight rounds per minute. By altering the gun elevation is it possible to put four shells on target simultaneously.

Turret
The enlarged turret of the PLZ-52 houses an autoloader derived from a Russian design, replacing the unsatisfactory PLZ-45 version.

to amphibious landing ships or even warships. Whereas a battleship of World War II era might be impervious to a mere 155mm (6.1in) shell, modern ships are not designed to shrug off multiple hits by powerful warheads.

Coastal artillery

Conversely, a battery of artillery vehicles on the shore – or some distance inland – is far more flexible than a fort or coastal artillery emplacement, and virtually impossible to hit even if it can be located. A 155mm (6.1in) self-propelled artillery piece is likely to outrange the gun or guns of a warship, while missiles have to find a mobile target concealed among landward clutter. In a changing world, coastal artillery appears to have gained a new lease of life, protecting potential invasion beaches and the approaches to cities, harbours and installations. Narrow waters such as straits can also be denied to an enemy navy in this manner – few powers will risk running expensive warships through an area where concealed enemy vehicles can target them at short range from positions on the shoreline.

Export model

The PLZ-52 is sometimes cited as being an export version of the PLZ-05, and on other occasions as a separate vehicle. The PLZ-04 is an export version of the PLZ-05, with a slightly longer 54-calibre gun barrel. It is in use with Kuwait and Saudi Arabia, both previous buyers of the preceding PLZ-45.

Without the support of a barrel rest, a long gun may deform under its own weight. Stresses are greatly increased when traversing uneven ground at speed.

PLL-09

The PLL-09 is one of the vehicles intended to replace the Type 89 self-propelled gun. The PLZ-07 fulfils the tracked SPG role, while the PLL-09 brings 122mm (4.8in) fire support to formations using wheeled vehicles.

These light formations typically use 8x8 vehicles from the same family, simplifying maintenance and ensuring that all vehicles are interoperable.

The PLL-09 is based on the same chassis as the ZBL-09 Snow Leopard APC. It is fully amphibious and is propelled in water by twin waterjets. On land its 328kW (440hp) engine gives a maximum speed of 100km/h (62mph), while rough terrain performance is comparable with many armoured vehicles. Its weapon is the PL96 howitzer, an improved version of the venerable Type 86 that arms many older Chinese vehicles.

The gun is directed by an advanced fire control system that is probably linked into an integrated battlefield management system, enabling a light formation to operate much more closely than previous generations of combat vehicles would have been able to.

A group of artillery vehicles, even with differing calibres and types of weapon, can saturate a target by tying all their fire control systems together. By varying the gun elevation to give specific flight times, a vehicle equipped with an automatic or semiautomatic loader can put multiple shells on target at the same time; a group of artillery weapons all doing so at the same time can create an overwhelming salvo.

The concept behind this technique is nothing new; time-on-target

PLL-09

Date 2009
Number Produced Not Known
Weight 21 tonnes (20.6 tons)
Length (with gun) 8m (26ft 3in)
Width 3m (9ft 10in)
Height 3m (9ft 10in)
Engine Deutz BF6M1015C diesel 328kW (440hp)
Maximum Road Speed 100km/h (62mph)
Crew 4–5
Armour Not Known
Main Armament 122mm (4.8in) howitzer
Main Gun Ammunition Stowed 30

PLL-09

The tracked PLZ-07 can go places the PLL-09 cannot, but the wheeled version has greater strategic mobility and a higher speed over the terrain it can traverse.

artillery barrages were used in World War II. However, with the advent of computerized targeting and coordination between vehicles there is less variation in the arrival times of ordnance, greater concentration in the target area and less chance that an error will alert the enemy with an early shell. The first seconds of an artillery attack are usually the most destructive, and vehicles like the PLL-09 are equipped to make the most of those seconds.

Advanced gunlaying has other applications, such as hitting enemy positions just ahead of an assault to keep hostiles under cover for as long as possible, or suppressing anti-armour weapons with precisely targeted artillery fire. A retreat can also be covered in the same way, enabling a light force to strike then break contact and retire. This can be important for reconnaissance assets that encounter a major enemy concentration.

PLZ-07

Developed as a replacement for the Type 70-I, Type 85 and Type 89 self-propelled guns, the PLZ-07 is built around a 122mm (4.8in) howitzer.

The gun is supported by a semi-automatic loader and aimed by advanced electronic systems, increasing rate of fire and accuracy compared to previous weapons of this type. It can deliver all Chinese 122mm (4.8in) munitions including extended-range and guided shells, and is presumably also compatible with Russian ammunition. This would be an advantage in the export marketplace, though as yet an export version has not appeared. The same ammunition is used in the wheeled PLL-09 self-propelled gun and a truck-mounted howitzer designated PCL-09.

The chassis of the PLZ-07 was developed from the ZBD-97 (or Type 97) IFV. Its more powerful engine gives it greater cross-country mobility than preceding self-propelled gun designs, though it is not capable of amphibious operations. It is not clear whether or not a flotation kit is available for the standard PLZ-07, but it seems likely since an amphibious version exists.

The amphibious PLZ-07B is used by marine formations. It is very slightly heavier and incorporates a pair of waterjets for use when swimming. The PLZ-07B is not apparently designed for high-speed transits to shore from distant amphibious warfare ships, and would thus follow the assault formations ashore once a foothold was gained.

Russian calibre ammunition

The PLZ-07 and PLZ-07B use 122mm (4.8in) ammunition, which is originally a Russian calibre, whereas many other weapon systems have gone over to NATO standard ammunition. In some ways this reflects the unique nature of the Chinese military – influenced by both the Cold War powerblocs but capable of charting its own path. There is no contradiction inherent in the use of 'western' and 'eastern' calibres in Chinese weaponry, since ammunition can be locally produced or sourced as needed, and the choice of calibres is internally consistent. If divisional artillery weapons use a Russian-

PLZ-07 Self-Propelled Gun
Date 2009
Number Produced 200+
Weight 24.5 tonnes (24.1 tons)
Length (with gun) 6.6m (21ft 8in)
Width 3.28m (10ft 9in)
Height 2.5m (8ft 2.5in)
Engine BF8M1015CP diesel 440kW (590hp)
Maximum Road Speed 60–65km/h (37–40mph)
Crew 5
Armour Not Known
Main Armament 122mm (4.8in) howitzer
Main Gun Ammunition Stowed Not Known

derived 122mm (4.8in) system and heavier artillery uses a NATO 155mm (6.1in) calibre, there is no more a supply problem or inherent inefficiency than if any other two distinct calibres were used for different weapons.

Amphibious capability

In addition to a willingness to mix design concepts and manufacturing standards, the Chinese military has developed other unique characteristics. Few nations produce fully amphibious armoured vehicles, and amphibious tanks are rare indeed. Yet China perceives the need and has designs available for export to other nations that also require them. Fully amphibious vehicles are highly useful to countries with a great deal of wetlands or large wilderness areas with few roads and many watercourses, and it may be that vehicles such as the PLZ-07B will find favour there.

Chinese vehicles and personnel are often equipped with digital camouflage patterns, sometimes in unusual colour schemes. Digital camouflage is the subject of much debate at present, with some research suggesting it is more effective than traditional camouflage patterns and some experts contesting this view. The fact that China has invested so heavily in cutting-edge and even experimental concepts of this sort is an indicator of how far military technology has advanced in the roughly 70 years since the end of the Chinese Civil War.

Gone are the crude copies of obsolete Russian tanks and armoured personnel carriers, although their legacy can still be discerned here and there. From a hodgepodge of legacy equipment, the Chinese armed forces have been transformed into a first-class modern fighting force equipped with hardware manufactured to their own specifications.

Chinese sources are notoriously cagey about the capabilities of their equipment, but as their designs increasingly appear on the export market it is becoming clear that third-generation Chinese tanks and armoured vehicles are as good as anything on offer elsewhere.

PLZ-07 Self-Propelled Gun

Chinese divisional artillery appears to be standardised at 122mm (4.8in), with other weapons concentrated in specialist artillery formations.

Abbreviations and Glossary

Ammunition Abbreviation	Meaning
AP	Armour Piercing
API-T	Armour Piercing Incendiary (Tracer)
AP-T	Armour Piercing (Tracer)
APBC	Armoured Piercing Ballistic Capped
APDS	Armour Piercing Discarding Sabot
APFSDS	Armour Piercing Fin Stabilized Discarding Sabot
APFSDS-DU	Armour Piercing Fin Stabilized Discarding Sabot (Depleted Uranium)
APFSDS-T	Armour Piercing Fin Stabilized Discarding Sabot (Tungsten)
APHE	Armour Piercing High Explosive
APSE	Armour Piercing Special Effects
FRAG	Fragmentation
HE	High Explosive
HE-FRAG	High Explosive (Fragmentation)
HE-FRAG-FS	High Explosive Fragmentation (Fin Stabilized)
HE-FS	High Explosive (Fin Stabilized)
HEAT	High Explosive Anti-Tank
HEAT-FS	High Explosive Anti-Tank (Fin Stabilized)
HEAT-MP	High Explosive Anti-Tank (Multi Purpose)
HEAT-MP-T	High Explosive Anti-Tank (Multi Purpose) (Tracer)
HE-I	High Explosive (Incendiary)
HEI-T	High Explosive Incendiary (Tracer)
HEP	High Explosive Plastic
HESH	High Explosive Squash Head
HVAP	Hyper Velocity Armour Piercing
HVAPDS	Hyper Velocity Armour Piercing Discarding Sabot
MDZ	Modernized Instantaneous High Explosive Incendiary
Shrapnel-FS	Shrapnel (Fin Stabilized)

Abbreviation	Meaning	Description
ACV	Armoured Command Vehicle	Armoured infantry-carrying vehicle that houses communication/command/control/computer/ISTAR systems (C4ISTAR).
AFV	Armoured Fighting Vehicle	Any combat-oriented armoured vehicle, including MBTs, ACVs, AIFVs, APCs and ARVs.
AIFV	Armoured Infantry Fighting Vehicle	Armoured infantry-carrying vehicle that sports significant firepower (usually in the 25–75mm/1–3in calibre range) that also often has pistol ports for the transported infantry to fire through when embussed.
APC	Armoured Personnel Carrier	Armoured infantry-carrying vehicle that sports modest firepower (such as a 7.62–12.5mm/0.3–0.5in machine gun) and that typically lacks pistol ports for the infantry it transports.

APS	Active Protection System	AFV protection device that actively engages an enemy threat, for example a device that shoots out shrapnel to detonate an incoming ATGM when it is a few metres from the vehicle.
ARV	Armoured Reconnaissance Vehicle	High-mobility armoured vehicle, sometimes infantry-carrying, that is optimized for recce and scouting missions.
ATGM	Anti-tank Guided Missile	Tank-killing missile that can be fired from an AFV that is wire-guided to its target by its operator, usually located within the AFV.
BATW-ERA	Built-in Anti-tandem-Warhead-ERA	Advanced type of ERA block designed to stop penetration by tandem warheads, which are specifically intended to defeat ERA protection by having two sequenced charge detonations.
BM	Battle-Management	Dimension of military activity that pertains to the coordination and synchronization of individual tactical activity into a larger integrated scheme; see ISTAR.
CIWS	Close-in Weapons Systems	A short-range missile defence system, often used to protect naval vessels from close-in threats.
Dual-ERA	Dual-ERA	Variant type of Built-in Anti-tandem-Warhead-ERA.
ECM	Electronic/Electro-optical Counter-Measures	'Soft-kill' or passive AFV-mounted device that disrupts the laser designators, laser rangefinders and SACLOS-guidance used to direct an ATGM onto its target.
ERA	Explosive Reactive Armour	A modern addition to AFV protection, an ERA block contains an explosive element sandwich between steel plates that detonates outwards when hit by an incoming enemy round or missile, thus disrupting the forward movement of the threat through the AFV's armoured plates.
FCS	Fire-Control System	Ballistic computer linked to the vehicle's ISTAR elements (commander's/gunner's sights, rangefinders, periscopes, sensors and digitized networks) that controls an AFV's (particularly an MBT's) lethality.
FLIR	Forward-Looking Infra-Red	Advanced military camera that detects the infrared radiation associated with heat sources such as an enemy vehicle or soldier.
FSV	Fire Support Vehicle	High-lethality AIFV variant that mounts a large-calibre main armament, usually a 90–105mm (3.5–4.1in) cannon.
GPMG	General Purpose Machine Gun	Versatile machine gun that can be swiftly changed from the light (bipod-based) role to the tripod-based medium role.

GPS	Global Positioning System	Extremely accurate navigation system based around the use of orbital satellites.
ICBM	Intercontinental Ballistic Missile	A long-range (5500-km, 3400-mile) strategic missile system whose trajectory takes it beyond the Earth's atmosphere.
IED	Improvised Explosive Device	Any improvised bomb or mine containing an explosive charge and some detonation trigger, typically made by insurgents.
IR	Infra-Red	An AFV sight or camera that detects the infrared radiation produced by heat sources (enemy personnel or AFVs).
MBT	Main Battle Tank	Heavy, 40–70-tonne (39–69-ton) and very well-protected fully tracked AFV that features a turreted large-calibre main armament (such as a 100–125mm/3.9–4.9in cannon).
MRAP	Mine-Resistant, Ambush-Protected	A type of armoured troop-carrying vehicle that is heavily protected, especially against the threats posed by landmines or roadside IEDs.
NATO	North Atlantic Treaty Organisation	The liberal-democratic Western European/North American military alliance in opposition to the authoritarian Communist Warsaw Pact military alliance during the Cold War.
NBC	Nuclear, Biological and Chemical	The various non-conventional threats against which most modern AFVs are protected.
NBCR	Nuclear, Chemical, Biological, and Radiological	Alternate term for the various non-conventional threats against which most modern AFVs are protected.
NERA	Non-Explosive/Non-Energetic Reactive Armour	Add-on armoured block, like ERA, that has an inert energy-absorbing substance (such as rubber) sandwiched between two metals plates, rather than an explosive charge.
NLOS	Non-Line-of-Sight	Advanced direct-fire weapon munitions delivered by AFVs that possess such great range that they travel beyond the horizon, thus moving out of sight of the firer.
PGM	Precision-Guided Munitions	An aerially delivered missile that is accurately guided to its target, often through use of a laser designator.
RHA	Rolled Homogenous Armour	Standard protection for AFVs until the 1980s – solid sheets of single-metal plate, usually steel; increasingly replaced/augmented by composite laminated armour.
RPG	Rocket-Propelled Grenade	Grenade-like warhead fired by an enemy soldier delivered from a launcher fitted to a small arm such as an assault rifle.

RWS	Remote-Weapons Station (or Remotely-Controlled Weapons Station)	Unmanned turret-like structure fitted to the top of an AFV, which mounts auto-reloading firepower weapons that are controlled from inside the vehicle.
SACLOS	Semi-Automatic-Command-to-Line-of-Sight	Guidance system for an ATGM where the operator aims a sighting device (like a laser designator) at the target and the missile guides itself following the beam to the target.
SAM	Surface-to-Air-Missile	Ground-based missile system employed to destroy in-flight enemy platforms, particularly aircraft, helicopters and drones.
SHORAD	Short-range Air Defence	An anti-aircraft missile system designed to engage enemy threats at close range.
SLAT-A	Slat-Armour	Known also as bar-armour or cage-armour, a form of add-on AFV protection that fits metal frames to the outside of the vehicle's hull and turret, designed to help protect against enemy RPG or HEAT (chemical-energy shaped-charged) threats.
SPAAG	Self-propelled Anti-Aircraft Gun	An anti-aircraft gun mounted on an armoured chassis.
SPAAML	Self-propelled Anti-Aircraft Missile Launcher	An anti-aircraft missile system mounted on an armoured chassis, such as the 9K35 Strela-10SV.
SPG	Self-propelled Gun	An artillery piece mounted on an armoured chassis.
TCV	Troop Carrying vehicle	Alternate designation for an APC or AIFV (see also ICV) (qqv).
UAV	Unmanned Aerial Vehicles	Remotely controlled (from the ground) aerial platform for recce or strike roles (also known as a 'drone').

Deployment of PLA Ground Units

PLA Combat Units

Due to the PLA constantly restructuring its combat units – whether through downsizing, reorganizing or disbanding units – the characterization and number of units shown here is approximate as of 2017. Personnel numbers comprise group armies, marines and airborne, including subordinate units not listed in the table.

PLA COMBAT UNITS	
Group Armies/Corps	13
Combined Arms Brigades	78
Air Assault/Army Aviation Brigades	13
Artillery Brigades	15
Airborne Brigades	6
Marine Brigades	7
TOTALS	
Personnel	915,000
Tanks	7400
Artillery Guns	10,600

Opposite: The size of Chinese territory and the varying terrain within it requires not only large numbers of armoured vehicles but also a variety of capabilities – and this is without considering the possibility of amphibious forces and overseas deployments. Desirable as it might be to equip all formations with the same first-class tanks and armoured vehicles, it is simply not practicable or necessary, given the variety of environments Chinese forces operate within.

Sources for pages 122–123:
Office of the Secretary of Defense. *Annual Report to Congress: Military and Security Developments Involving the People's Republic of China* (pages 27 and 123).

Index

References to illustrations are in *italics*.

Q

R

S

T

Picture Credits

81.cn/Wang Ning: 88
Alamy/ITAR-TASS: 56–66 all, 87, 94
Amber Books/Art-Tech: 22–34 all, 107
Cody Images: 13, 16, 18, 39
Getty Images: 6 (TASS/Sergei Bobylev), 8 (AFP/Greg Baker), 42 (AFP/Amir Quereshi), 48 (TASS/Sergei Bobylev), 51 (AFP/Greg Baker)
Chamal Pathirana: 70 & 77 (CC BY-SA 3.0)
Shutterstock: 7 (Boris Dianov), 45 (ID1974)
Tyg728: 80 & 101 (CC BY-SA 4.0)

Artworks:
All artworks courtesy of **Tank Encyclopedia**, except for the following:
Amber Books/Art-Tech: 100, 108